Praise for *The Berry Grow*

Whether you're looking to grow for your family, o anywhere in between), this resource is absolutely sta insights that you can utilize. Blake has done an amaz an enormous amount of information into an easy to follow resource that anyone can use.

—Rob Avis, PEng, co-author, *Building Your Permaculture Property* and *Essential Rainwater Harvesting*

Right at the beginning of his book Blake Cothron makes a good case for growing berries and small fruits. It's the flavor of these fruits when picked ripe. Or, if you're an aspiring market gardener, it's the returns from these jewels. From there, Blake covers everything from soup to nuts, figuratively, in growing these fruits in various settings, for home use or markets. A number of the fruits he mentions are not yet well-known, but are worth growing.

—Lee Reich, PhD, author, *Growing Figs in Cold Climates* and *The Ever Curious Gardener*

The Berry Grower is a terrific resource for anyone getting into small fruit production or those looking to diversify their crops. It is loaded with helpful background, tips, tricks, and recommendations that will benefit small fruit growers of any scale or level of experience!

—Dan Dalton, PASA Sustainable Agriculture

Blake uses 20 years' experience growing berries and other small fruits to provide valuable practical, effective, up-to-date information (including addressing climate chaos) to encourage more localized and resilient organic food production, garden by garden. Small fruits are a good place to start because they bring harvests 6–12 months after planting and the demand is almost infinite. Here's trustworthy information about planting and growing blackberries, red and black raspberries, strawberries, currants, figs, gooseberries, juneberries, muscadine grapes, blueberries, passionfruit, and—surprise—tomatoes, with shorter profiles of almost 20 other lesser known small fruits. That's a big diversity of crops in one book!

—Pam Dawling, author, *The Year-Round Hoophouse* and *Sustainable Market Farming*

I've cracked open your book and like what I'm seeing so far!

—The late, great Michael Phillips, author, *The Holistic Orchard* and *The Apple Grower*

THE BERRY GROWER

SMALL SCALE ORGANIC FRUIT PRODUCTION IN THE 21ST CENTURY

BLAKE COTHRON

Printed in Canada. First printing May 2022.

Any other inquiries can be directed by mail to:
New Society Publishers P.O. Box 189,
Gabriola Island, BC V0R 1X0, Canada
(250) 247-9737

LIBRARY AND ARCHIVES CANADA CATALOGUING IN PUBLICATION

Title: The berry grower : small scale organic fruit production
in the 21st century / Blake Cothron.

Names: Cothron, Blake, 1985– author.

Description: Includes bibliographical references and index.

Identifiers: Canadiana (print) 20220200742 | Canadiana (ebook) 20220200769 |
ISBN 9780865719651 (softcover) | ISBN 9781550927580 (PDF) |
ISBN 9781771423540 (EPUB)

Subjects: LCSH: Fruit-culture. | LCSH: Berries. | LCSH: Organic farming.

Classification: LCC SB357.24 .C68 2022 | DDC 634b

Funded by the Government of Canada | Financé par le gouvernement du Canada

New Society Publishers' mission is to publish books that contribute in fundamental ways to building an ecologically sustainable and just society, and to do so with the least possible impact on the environment, in a manner that models this vision.

Dedicated to Bhagavan Sri Krishna
whose Name is the path back home for all Jivas,
to Paramahamsa Vishwananda,
and to Mother Bhumi,
the shining jewel
in the vast darkness.

Contents

Preface

We live on a planet in crisis. As I was writing this book, I deeply contemplated what contribution I could make that would, in its own small way, be of positive assistance in this time of imminent transition. I firmly believe that organic farming, especially with an accompanying shift to localized and more resilient food production, has to happen on a global scale if our civilization and societies are to survive and thrive into the next century and beyond. I feel that education and empowerment through practical, effective knowledge and training is crucial to initiating this changeover.

So, I decided to write a book to help people grow healthy local food—specifically, organic fruit. I didn't want to write a generic, flashy, marginally useful coffee table book about "how to grow berries." Those already exist. So, what could I do to set this book apart and make it most useful?

It's my goal with this book to share innovative strategies for small fruit growing, with pertinent 21st Century information to help progress the organic farming, urban farming, and local food movement. The focus of the local food movement has been intensely about vegetables, which are rapidly grown and profitable. I'm hoping to help some adventurous growers venture into small fruit growing as well, which has many advantages.

This book is a modern strategy guide to how small fruits can work for you in your small or micro farming operation—or in your own backyard business. If you're just looking to grow organic small fruits for your family, this book will help you. I think backyard growing can not only help create abundance, extra wealth, and food security but is often a perfect starting point to train yourself to prepare to upscale to market growing.

> The greatest change we need to make is from consumption to production, even if on a small scale, in our own gardens. If only 10% of us do this, there is enough for everyone. Hence the futility of revolutionaries who have no gardens, who depend on the very system they attack, and who produce words and bullets, not food and shelter.
>
> —Bill Mollison, the late Permaculture founder and teacher

How can we get to that 10% or even 25%? Will it take disasters, food shortages, pandemics, war, and pestilence to force us out of couch-lock complacency and into our backyards? Will thousands of deaths from *E. Coli* contaminated strawberries finally convince us to relocate from Costco to the corner farmers market to seek out our fresh produce? Will our governments jump in and save us with truckloads of food? Where would all that food come from? Will we grow it on Mars in hydroponic greenhouses?

The needs of the day are met with solutions that are utterly practical, grounded, and highly efficient. We must utilize ways to grow food and to address other basic needs, while doing minimal harm to the Earth, our dear Mother. Moving beyond "minimal harm" we come to regeneration, in which resources are created, used, and reused in closed loops, so efficiently that more land is put back into habitat and we can begin to heal the harms that have been done.

The Earth is extremely patient and has tolerated a great deal of harm, but that damage is catching up to Her and the systems we rely on are reaching critical breaking points. How long will semi-truck loads of California fruit criss-cross every nook of the country to stock every market in America? How long will the devastated soils of California continue to produce? How long will the mountains flow with sufficient melted snow pack to stock the rivers that irrigate those crops? What happens when they no longer do so?

It's high time we transition from reliance on trucked-in food and put intense emphasis on local production. It's going to take educated and inspired people doing it. We've made a lot of progress in the last 20+ years in this regard but much more work remains to be done. I'm always amazed when I go into the local health food stores, even in major cities, and still see almost no local produce. The niche is wide open, but where are those growers who are ready to take it on?

Despite a lack of enterprising local growers in many places, there continues to be an exponentially growing interest in local food production itself, including fruit. This includes edible perennial landscaping in urban city parks, mini-orchards in suburban backyards, homesteading and self-sufficiency, as well as organic market farms looking to diversify their product line and increase customer draw.

There is intense interest right now in growing tree fruits: apples, pears, citrus, pawpaws, etc. However, this book is focused primarily on the sometimes undervalued "small" fruits: berries, figs, grapes, and even tomatoes. These are primarily produced on shrubs, bushes, or vines and therefore have many strategic advantages over growing tree fruits. For instance: small fruits exhibit comparatively rapid yields compared to tree fruits: only six months to a year for many small fruits to get into production, as opposed to 2–5 years or more for tree fruits to come into useful production. Plus, the square footage needs of a tree (as well as the years to production) is sometimes more than an urbanite or backyard grower can afford. Trees can shade out gardens or solar panels that need full sun, while small fruits can fit almost anywhere, even in containers on a balcony.

This is not to say small fruits are superior or that tree fruits aren't important and vital for feeding the world—of course they are. Both are a specialty of mine and something I grow plenty of on our organic research farm in Kentucky. But this book is focusing on small fruits for (often) small spaces, on a small planet facing very big challenges.

As you research fruit growing you will discover the limitations of what is available. Many older fruit growing books are nearly obsolete or at best teaching inefficient and outdated techniques. That includes the advice, fruit varieties, and strategies for growing fruit offered in fruit growing books published in the 1970s, 1980s, or even up to the early 2000s. A lot has changed. New, more efficient techniques, tools, and strategies are out there. The market has undergone huge shifts. New and ever-arriving invasive insects are a constant threat. How can you best strategize for that? Many old, popular cultivars are no longer popular or even available now or have been surpassed by new, superior ones. Which current ones are vastly superior? Climate change has changed the game. How can you best prepare for all these shifts and challenges?

I've spent over 20 years successfully growing organic fruit and vegetables across a number of varied locations in the USA and nearly 10 years ago created and currently co-operate a successful commercial organic plant nursery. On our farm we grow over 15 species of organic small fruits. I've marketed organic fruit and hundreds of types of horticultural products, and spent hundreds of hours researching growing organic fruit. I have also taught a number of in-person courses and all that experience has gone into this book.

Throughout the book, I've aimed for meaningful, highly accurate information gleaned from my experiences, both professional and as a lifelong fruit explorer and grower. To include an array of useful perspectives and strategies, I've included valuable stories and first-hand experiences from many other very successful growers across the USA. From there you will need to seek out local growers, grower's associations, regional publications, online grower communities, and your local agricultural extension offices in order to zero in on the most accurate locally-oriented information.

Finally, I believe we can do it. I believe in the utter resiliency, intelligence, and adaptability of the human being, as well as the grace of the Divine, both of which I believe will be needed for us to survive this global transition in consciousness, from exploitation and devastation to integrity, unity, and harmony.

PART 1

The Basics

1

Why Small Scale, Small Fruits?

The fruits of Paradise dangling down from green leafy bowers, so heavily laden they nearly touch the ground; flowers humming with honeybees and ripe fruit dripping sweet nectar in the sun.... There is something about *fruit* that conjures up these archetypal images, in ways that kale and radishes simply do not. Ancient Vedic and other histories tell us the Earth used to hand us Her bounty in much, much greater generosity, with little to no labor on our part, and no doubt we accepted it much more gracefully.[1] Today we must toil and sweat to beg Her fruits from those somewhat laden branches.[2]

Why small fruit growing?

Organic fruit is delicious, healthy, and *fruit sells*! Through a modest planting of small fruits such as berries, figs, and tomatoes, you can not only feed yourself but, if you're a market grower, you can stack your existing market table with piles of high-value colorful fruit. This alone can provide a strong customer draw and set you apart from other market growers. Strategically adding a small fruit planting to your market operation will increase your overall labor very little but will bring many benefits.

Likewise, converting your backyard from grass into fruit production is also a very rewarding process. As well as harvesting household fruit, there is also the viable possibility of marketing excess fruit—as well as the seeds, cuttings, and fruit plants themselves. These products are high value and in demand and can often be harvested from the same planting.

Although, in the first quarter of the 21st Century, widespread availability of USDA Certified Organic fruit in supermarkets is now fairly common, the quality is just not the same as locally grown and yet the price tag is still very high. Demand for high quality, ripe, hand-picked, local organic fruit is rising exponentially. This wide-open niche is there for the skilled, strategic market grower to fulfill.

However, I must advise some caution. Fruit farming is not something I would recommend most people take on as a full-time occupation. This book is not about becoming a full-time fruit farmer, nor is it about farming organic fruit on 50 acres. It's about equipping yourself with practical knowledge so you can understand how adding small fruits or berries to your market farming operation, starting a micro-growing operation on an acre or less, or just growing in your backyard for fun and profit can be done successfully. I have placed home and market growing recommendations in Part 2, after each individual fruit is detailed. The techniques and fruits described will work just as well in a backyard setting as on the small farm and will have you set up for success no matter your scale or purpose with growing small fruits.

How and where to start?

You can start growing small fruits just about anywhere there is good sunlight and a little land. Even containers on a sunny balcony can be used for tomatoes, passionfruit, raspberries, and more. The average backyard can produce an amazing abundance of fruit, with enough extra to sell for a side income. It's up to you to take the initiative and get started. If you have a small unused parcel on an existing market farm, you can start there.

I'll share a small fruit marketing story from 2014–2017, when my wife and I were vegetable market farming in Appalachian Kentucky. In March we planted about 150' (46 m) of row of fall (everbearing) Caroline red raspberries, divided into six 25' rows. Six months later we started bringing ½ pint clamshells of raspberries to our quiet, small-town farmers market with decent organic food demand, in Berea, KY. We marketed them for $5 each, which was the current grocery store price. We could not bring enough. Every week we sold about $150–200 in berries, which at the time was a substantial boost to our overall weekly income. We also propagated and sold the plants, bringing in additional income. Picking the berries only took about 4–5 hours a week, which we did on the two days leading up to Saturday market (Thursday, Friday, and also Saturday morning). We carefully graded the berries by hand and chilled them immediately after harvest. We only took to market the A+ and some B+ grade berries (based on size and appearance). Picking earlier than three days before market would have been too long of a storage time and could have risked the berries molding. The other days of the week (Sunday–Wednesday) we picked the berries and sold those through other outlets (a health food store and our home delivery service). This brought in even more income. Also, note that the raspberries were *not yet even close to their peak production*; this was only season one and they were being grown on marginal ground. Had they been on fertile soil in their peak production, the yields would have been 2–4 times heavier.

Overall, it was a tiny expenditure of land (150' of row), capital (about $150 in T-posts and plants), and labor (4–5 hours a week picking and packing). And yet, during berry season (July–September) the raspberries boosted our income by some $600–800 per month. Not bad! There's no way a large commercial farm could get numbers like that on 150' of berries. A small planting like this could easily fit in many backyards and, I'm sure, a lot of you out there would enjoy eating homegrown raspberries and earning $800 a month in additional income throughout much of the summer, while providing organic fruit for your local community. Before we move into how to do it, we first should understand a few challenges to growing fruit and why small fruits make big sense.

Understanding current climate challenges

With climate change, you can be certain of one thing: uncertainty. Continually altering weather patterns make it necessary to be ready to deal with unexpected occurrences. These include lack of (or too much) rain, colder (or hotter) than average temperatures, as well as altered spring warm-up times (often earlier). Below are some detailed accounts of how climate change events have affected our plantings.

In the winter of 2015–16 we experienced what some refer to as a "polar vortex" event, wherein our winter lows were about 10°F (6°C) colder than the extreme low for USDA zone 6: –10°F (–23°C) extreme low. Wind chill took it below –10°F. This affected marginally cold-hardy plants and some growers lost trees and plants.

In the autumn of 2019 to spring of 2020, we witnessed unprecedented climatic things happen in our local area. First, we had a severe late summer to early autumn drought wherein there was very little to no rain for about 8 or 9 weeks, with temperatures in the 90s (32–37 in °C) almost the entire time, and intense sun. That is not common for KY, although it had been recorded previously. Many half-century-old pine trees turned permanently brown, and many small tree saplings in front yards died also. By late September rains returned and then we had an unusually warm autumn. This delayed the hardening off (lignification or production of wood/bark) of many of our fruit shrubs and trees, which stayed very green into October. Then, all of a sudden, it dropped to 20°F (–7°C) one night in mid-October. This shocked and damaged these plants as they had not begun the process of hibernation, nor lignified their wood, in preparation for freezing weather. Winter set in shortly after.

Winter that year was mild, never going below about 10°F (–12°C: USDA zone 8 conditions, whereas it's normally Zone 6) with daytime temps around 40–50°F (4–10°C). Spring came on very early, with early March warming up to 70°F (21°C). Very pleasant, but far too early for it to be that warm every day in our region. With the early warm up, fruit trees and bushes woke up early. Our mulberries, jujubes, grapes, peaches, pawpaws, pears, apples, and most everything else sprouted leaves and flowered a month early. Then... BAM. Spring freeze down to 26°F (–3°C) in late March. Everything melted. The pears, peaches, and apples lost all their fruit, pawpaw flowers were fried, mulberries and jujubes were toast. After that, the plants started to *slowly* recover and more blooms came out on the pawpaws, and fresh green leaves began sprouting again on the fruit trees and shrubs. Then, BAM. A late mid-April freeze down to 28°F (–2°C). Everything got fried *yet again.*

This second time, it proved too much. The deep stress of it all to the plants led to our first-ever infestation of Asian ambrosia beetles (*Xylosandrus crassiusculus*). This opportunistic little beetle takes the life out of struggling and unhealthy trees and shrubs of many kinds, including pawpaws. It senses ethanol,[3] which is a byproduct of trees that are stressed out. It's as if the stressed trees were waving a bright white flag saying "I'm weak and vulnerable and ill-adapted," and the ambrosia beetle is death's scythe as a response. Many fruit trees died in our local region, and on our farm we lost dozens of specimens. This devastating weather pattern affected much of the southeast, but Kentucky was hit especially hard.

Small fruits to the rescue

However, we noted which fruiting plants were *essentially unaffected* by all of this climatic irregularity: the blackberries, red and black raspberries, strawberries, *aronia*, figs, gooseberries, juneberries, blueberries, passionfruit, honeyberries; *all the small fruits were totally fine! Hmm...*

We observed all season that despite the late freezes, even during the time of full bloom for many of them, all the small fruit plants succeeded despite the extreme, unstable weather. We had amazing harvests of perfect blackberries, strawberries, raspberries, *aronia*, passionfruit, gooseberries, and more. As a positive side effect, we also had drastically reduced populations of stinkbugs, June beetles (*Cotinis nitida*), and Japanese beetles, all typically major fruit pests.

Around this time, I contacted a local fruit orchard about an 80-minute drive north of us, and I inquired as to the status of their crops. I was informed there would be *no apples, no pears, and no peaches in 2020.* Those fruits are their main crops and income drivers. What would they have available? Blackberries, raspberries, blueberries, and squash: *small fruits, to the rescue. Hmm...*

All of this gave me an opportunity to observe the resiliency of all these plants, and collect data. Everything in farming and gardening can be seen as feedback, if you're listening and observant. Remember that.

Once again in 2021, we experienced a similar yet much milder version of 2020 spring weather. Early warm temperatures and then late freezes once again destroyed most of our tree crops. No pears, almost zero apples and pawpaws. Thankfully our other tree fruits stayed dormant longer this time and were unaffected (persimmons, jujube). However, once again the small fruits did great and were apparently completely unaffected.

Why micro fruit farming?

There are different ways to approach this question. Are you already a market farmer or do you currently work 9–5 in an office, possibly under fluorescent lights? If you're already a farmer and you're asking this question, you're likely looking at how to make your operation more efficient, reducing costs and increasing yields, while introducing new products. Or, you could be an organic farmer who is (possibly) overwhelmed and looking to downsize your operation to a more human scale. You may be facing the potential of losing your land or a land lease will soon be ending. Maybe you're inheriting the conventional family farm and you want to make it work organically. You may be exploring transitioning to organic farming, or taking on organic farming as a career choice (or career shift). Maybe you just want to transition your suburban lawn into organic food and cash.

If you're new to farming (maybe you're that office worker mentioned earlier) and the sunshine, soil, and sky are calling you home, then micro farming is a great place to start (and maybe stay). Basically, micro farming is farming on a plot that is 1⁄16 of an acre up to a few acres. For perspective, most modern US suburban lots range from 1⁄16 acre to about ¼ acre. Older suburban lots in larger cities, divided in the 1950s–1970s were often much larger, often around ½ an acre to 1 acre in size.

I remember those large suburban lots from when I was growing up in Louisville, KY. Specifically, I very fondly remember the abundance of elderly World War II veterans who intensively cropped their large backyards for household self-sufficiency, having turned them into productive mini-homesteads. Although not usually or completely organic, they were growing their own fruits and vegetables. They often sold or generously gave away the excess, they saved their own seeds, and their wives canned enough for winter, all from their ¼–½ acre suburban back lot. Yet now, with the recent disappearance of nearly all of those quiet, humble heroes, so too have their abundant urban homesteads quietly reverted back to lawns.

Producing on this small scale is sometimes called *market farming* or even just *gardening*, but *micro farming* or, when in the city, *urban farming* better describes the intent of this kind of small, yet often highly effective operation and defines that you're selling product. It's the *human scale farm* that worked effectively for millenium after millenium before the advent of industrial-era mechanized farming and agricultural chemicals. Beginning in the late 1800s farming was able to displace people power, going beyond human scale to an artificial industrial scale. Entire books have been written on this tragic turn of events with its profound implications and global repercussions.

A major draw of micro farming is that, depending on the type of operation, it's reasonable that one physically fit, youthful person can farm fresh produce on ¼–½ acre of land, sometimes more, by themselves, given proper tools, training, strategies, and experience—and without a riding tractor. Any size more than that rapidly starts to become overwhelming and unproductive, unless more hands are brought in. Two fit, trained people could handle about 1 acre of produce by hand, and maybe more if we're talking about growing fruit trees, berries, and shrubs.

This may not sound like much, but it's been showcased by many modern growers such as Elliot Coleman and, more recently, Curtis Stone and Jean-Martin Fortier, that this is not only efficient and highly productive, but also potentially highly lucrative. There are reports of $20–100k+ *per acre* being earned doing skilled micro vegetable farming, when combined with strategic marketing and a receptive, middle class or wealthy customer base. Compare that to the conventional farming forecast of

$500–2000 per acre for industrial monoculture crop returns (soy, corn, hay, etc.)

It's important to emphasize that a diversified organic farm (polyculture) does not necessarily have to be "micro" in size—they'll work on virtually any scale. But this book is focused on small scale, human powered, micro agriculture with minimal machines and inputs.

I've thus far mostly been referring to micro *vegetable farming*, which has quick yields (several months) and rapid returns, and high net profit potential. *Fruit growing* is different because yields take longer to come back. Berries and other small fruits take anywhere from 6 months to 2 years to produce profitable yields, fruit trees take 5+ years. However, the labor requirement for fruit is also much less intensive than vegetable farming. For instance, vegetable farming requires near constant replanting, maintenance, and harvesting in order to stay sufficiently productive and highly profitable. Also, weeding, marketing, strategizing, checking crop progress, pest control, etc. are all very time-consuming and laborious work. There is all of that involved in small fruit growing but in general it is much less intensive, and is concentrated into certain events, mostly as the fruit ripens.

Fruit often ripens all at once or within a few weeks, requiring only a brief harvest period, unlike greens or cucumbers which must be picked several times per week for months at a time, or entire seasons. *This aspect makes micro farming with fruit attractive to people looking to farm part-time* while keeping a part-time (or full-time) job, or retired people looking to stay occupied but not slammed with intensive vegetable farming and bustling, demanding farmer's markets. Fruit plantings can also be added into an existing micro farming produce operation to add diversity, interest, and income.

Micro farming is suitable for people on a low or fixed income, as it does not necessitate the purchase of large acreages of land on which to farm, or expensive farm machinery such as tractors, or huge investments like barns. Micro farming can be done in a large sunny backyard, on other people's land, leased land, etc. You will need to invest a modest amount of upfront capital to get started: quality tools, plant material, fertilizers, and some basic, very affordable equipment (such as trellis posts, bird nets, or irrigation lines), as will be covered in detail later on. But it's reasonable to say, if land is already in possession (or leased, etc.) than one could get into micro farming fruit for a few hundred to a thousand dollars, depending on scale, and maybe much less if you have more time than money and plenty of patience. We'll be exploring this in detail.

Limitations of small fruit growing

It's important to understand the limitations of growing small fruit versus vegetables, for profit. Intensive vegetable production, per acre, is likely always going to become profitable faster and, in general, will be more profitable than growing small fruits. Vegetables produce more sellable mass than small fruits and mature rapidly. Vegetables can be rotated and grown year-round, or at least most of the year, while small fruits are usually only harvested once per year or at most six months of the year. Vegetables are also easier to harvest, process for market, and transport. People and restaurants spend more of their food budget on vegetables than fruit.

This being said, small fruits are still a high-value crop and can *enhance your existing market operation* by adding interest and diversity, and could serve as a niche crop for an enterprising grower as well. Even in your own backyard, you could potentially make thousands per year converting your lawn (or neighbors' lawns) into high value berries, figs, or tomatoes, etc.

The labor to establish a small fruit planting is a one-time, yet long-term proposition, whereas annual vegetables must be continually replaced and replanted every few months. Small fruits are less ongoing labor, are perennially productive, are less common and higher value (per pound) than vegetables in most local marketplaces.

Maximize profits by being the workforce

One of the factors that makes micro and small farming potentially so profitable is that it's "human scale"—powered by people. That allows you, as the owner/operator, to perform most (or even all) of the labor yourself. Therefore, you don't have to pay anybody for labor, which is typically the largest annual small business expense by far.

University publications on crop profit potentials always seem to assume you're hiring all the labor out. Why? Hiring labor out is sometimes necessary, but can really eat away at any farm profits pretty fast. It's best to size the operation, at least when starting out, so that you (or you and your family/partner/etc.) can do 90–100% of the labor yourselves. This can be tough, sometimes extremely demanding, but is better than working a 9–5 job you don't like with a boss and coworkers you'd rather not work with. As a bit of personal advice, I can tell you it's usually not the best idea to enlist your girlfriend or boyfriend into your small business as free labor, and only slightly better to do this with your spouse. It may be fun and enlivening at first, but in the long run it's a great way to cause a lot of deep stress on the relationship. A family dedicated to farming or running a family business, however, is a different story, yet not very easy either.

The future of small and micro farming

Micro farming is not a passing fad or something "real" farmers don't do. It will absolutely be a major part of the future of global food production and farming. Micro farming will gradually take precedence as the insurmountable burdens of unpayable farm debts, exorbitant mega-equipment costs, increasing age of the average farmer, fluctuating and unstable global markets, soil erosion, water shortages, super weeds and super insects/diseases/viruses gradually bankrupts and cripples the Industrial model of extreme input, mega-sized, mono-crop farming. Micro and small farming is the model many in third-world countries and less affluent areas are taking on, reverting to (or modernizing to), as they see the Industrial mega-farm model already failing them and their neighbors. Take part in the future of farming while feeding local people and earning a right-livelihood—right where you are.

2

21st Century Strategic Planning

Planning it out

To start, you will need to have a conducive, suitable site for growing fruit. It needs to have the following basic parameters:

1. Full sun (strong, direct sun exposure) for 8 hours minimum; 10+ hours a day is best. It must be direct sun and not filtered or obstructed by trees, etc.
2. Good drainage of water and cold air (ascertained through observation).
3. Low enough water table (at least 18–24" (46–61 cm) below ground for small fruits, 3–4' (0.9–1.2 m) for grapes and trees). Test this by digging a hole on the site at least a few days after a rain (digging with a post hole digger is ideal). If you hit water, you are likely encountering the water table. Measure how far down it is. If you dig down 2 ft. (0.6 m) and do not encounter water, the water table is likely low enough for small fruits.
4. Some protection from wind and access to irrigation water.
5. Nearly level or sloping ground is best but slightly steeper slopes can also sometimes work.

If these parameters are not met on your site you need to find another more suitable site, or you could grow in large raised beds or containers. Raised beds are described in Chapter 4.

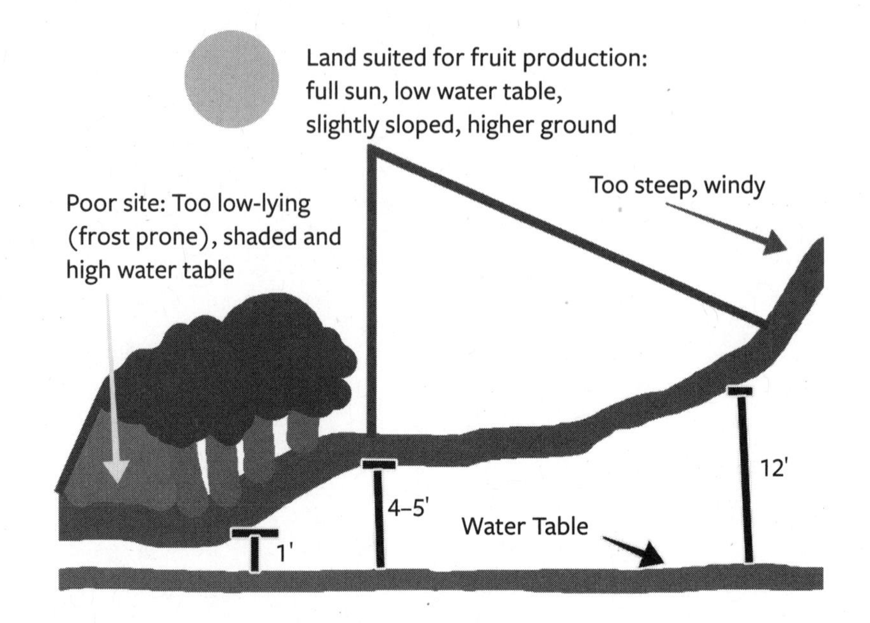

Start planning your fruit planting about six months to a year before you want to start. This way you can thoroughly research it for a few months, plan it out on paper, consult with regional authorities, perhaps visit other growers, and survey your customer base and local markets. It may take some time to do this, so do not rush it. It's best to have an order for plants ready and placed by autumn or early winter before nurseries start selling out for the following spring. Spring is the best time to plant most small fruits.

Although this can be an exciting endeavor, *never rush it*. I used to rush things. Through rushing and being impatient and overly passionate I met with lots and lots of unnecessary, expensive failures and very frustrating lessons. If you're *all of a sudden* inspired to plant strawberries and it's June, it's likely *just too late*. You don't want to get off to a bad start and have to dig up dead plants, wasting your time and resources, including emotional resources. Just wait till next year or next season and do it right, in the proper timing for your local area.

Species selection

The first step in the strategic planning process is to create a list of what you'd like to grow. Write down every small fruit you'd consider growing and why. Example: blackberries for the You-Pick, figs for side income, tomatoes as your main cash crop. What you're essentially doing here is contemplating what *species* you're interested in and why.

After that, you'll have to make sure the species you're interested in will not only *grow* but survive long term *and* produce adequate crops in your area. Just because you live in a compatible cold hardiness zone (say USDA zone 6) and blackberries "thrive in a zone 6 climate" does not mean you can grow blackberries successfully in your area.

For instance, here in Kentucky, cherries are fully cold hardy and *grow* just fine; the *trees* that is. However, in KY it's too wet, humid, and frosty in the spring for reliable, marketable cherry production and many years the whole crop rots on the tree. If you failed to do your research and planted "zone 6 compatible" sweet cherry trees in KY, you'd be facing eventual crop failure and be out of resources, time, space, etc.

So, *species selection is crucial* and is the first criteria to carefully research to see what is possible in your area. Also of vital importance is what species are *marketable and in demand* in your area. Let's look at how to best find answers to these questions.

How to research

In today's world, what passes for "research" often means googling and reading a few online articles from random sources and then coming to a speculative conclusion based on the culmination of that minimally sourced, questionably factual, and often highly biased information. That's not really research or, at least, not effective research. Recently there are many online articles about growing fruit that are actually a form of disguised advertising created simply to sell nursery plants or products and are often full of gross inaccuracies, false hopes, and oversimplifications.

If you want to be effective in your research you have to utilize a diversity of reliable sources of information, far beyond reading a few online articles. That means studying articles published by university agricultural departments, consulting with authorities on the topic, visiting people actually doing it, and studying quality books written by experts on the topic. Also, your local agricultural extension office is often a goldmine of information, which we will discuss in a moment. Through this holistic process of research you can most effectively be educated on what you're wanting to do.

Let's say you read online that 'Bluecrop' blueberries are "good for organic growing and high quality" and you're considering planting 100 of them. First, you must ascertain: is 'Bluecrop' recommended by your local agriculture extension office? Is it grown and marketed locally? Does the local You-Pick grow this cultivar? Does your local university say anything about it in their blueberry research papers? When is its approximate ripening date? Does it bloom too early for your local climate? Is it susceptible to local diseases and insects? Is it cold hardy enough for your area? You have to carefully research ALL these things from accurate, reliable sources, not nursery catalogs or random online articles making claims. Otherwise, you may invest a lot in something that does not pan out in your local area.

Begin with your local agricultural extension office

People make the big mistake of trusting their conclusion based on a rapid *google search* and not utilizing their own tax dollars by contacting their local agricultural extension office, the one that they did not even know existed nor know that they were financing. If you want to google, google your local ag extension office first! Most, if not all counties in the US and many areas of Canada, Australia, and Europe all have useful agricultural government departments open to the public.

Your local ag extension office can provide crucial and accurate answers to questions like, "*Will strawberries grow in my local area*?" and

"*What is the soil type on my site?*"[1] They will usually have available resources such as: locally based growing guides, marketing information, and lists of scientifically tested, proven productive cultivars for your specific area. Nice! This is far, far, better than googling "how to grow strawberries" and "best strawberry varieties for Texas" and trusting that some random article is totally accurate—*risky business!*

So, start your research with the ag extension office and generally you should trust their information. If they tell you "Strawberries will not grow in this county or are not able to be grown here profitably" then maybe forget about it, or do a very small trial of some promising cultivars and see what happens. They are basing that information on years of scientific field trials and research. Don't go into an endeavor thinking you're going to prove them all wrong or prevail against all the odds. I've seen a number of people attempt that sort of thing, and it usually fails. That's not the point. You're not out to do the impossible or beat the system. The point here is to produce healthy, local organic food and make right livelihood income, and your ag extension office is there to assist, not hold you down. In addition, they can provide soil testing and also applications and assistance with applying for *agricultural grants* that could be extremely helpful for you.

Seek out local growers and experts

Remember, just because you've "seen it growing" does not mean that a product is *marketable or in demand*, or will produce marketable yields in your area. You have to research all of this carefully and also talk to knowledgeable local people.

If the local office does not have pertinent information about what you're wanting to grow, the next best approach is to seek out local growers and try to consult with them, or visit their operation (perhaps posing as a curious customer) and try to get as much info as possible on what and how they do things, especially the cultivar names of their fruit species. Do an online search for local fruit farms, You-Picks, and growers, and talk to local co-ops and health food stores to see who they're buying from locally. They don't have to be organic growers to provide useful local info.

Also check into local grower seminars, grower associations, and workshops. These can be excellent places to meet other growers, network, and learn crucial information for growing in your local region. If no one in the local area is growing (or has even heard of) what you are considering, that is a "red flag" that you might be trying to swim upstream. Perhaps what you're thinking about is nearly impossible to grow locally or your local community has little to no interest in the product itself. This is useful feedback, especially for market growers, and should be carefully considered.

Local or regional botanical gardens are also great places to explore, as they very often have locally adapted fruits growing and usually have cultivar labels so you can see yourself what's thriving and how it's being grown. Contact them and inquire.

Your Desired Crop
↓
Ag Extension
Office Info + Feedback
↓
Seek out local growers and experts
↓
Survey Customers
Consider Local Preference
↓
Saturation and demand
of local markets and availability in
local stores and outlets
↓
Your practical ability
to produce and market it on
a meaningful scale

Survey your customers and observe the local market

If you are already a market farmer, survey your customers in person or online: "*What fruits would you be most likely to buy from us: strawberries, blueberries, blackberries, figs...*" Their feedback can help you target what would be most in demand, or just go with what is obviously in high demand. Also check local saturation levels at your market. If five other growers at your market are already bringing strawberries, you might want to bring something no one else has. If they are all conventional producers, you may be able to fulfill the demand for organic strawberries.

Investigate what small fruits the local health food stores or co-ops sell that are being imported from out of state. Could you grow that and provide it locally? Would they be interested in local organic product? Inquire.

Evaluating agricultural profit forecasts

Over years of studying many university and extension office documents on growing and marketing various crops, I've noted the dismally low and even somewhat pessimistic profit estimates they often calculate.

For instance, the University of Kentucky forecasted in 2017 that a 96' × 20' high tunnel growing tomatoes could annually earn you about $850 after expenses, including labor.[2] This abysmal number suggests a few things. First, the profit forecast is assuming the tomatoes are being sold for only $2 per pound, which is on the low side for direct marketing. Second, this number is not considering organic production and pricing. Third, for some reason these publications always seem to assume you're hiring all the labor out. If you're doing most or all of it yourself or within your family, then that profit estimate just got a lot bigger.

A local Amish family farm out here in central KY have a high tunnel nearly the same size and, according to what they told me, they're making about $3000–4000 annually on the conventional (non-organic) tomatoes produced in it, and that's selling wholesale for about $2/lb. on average. They perform all the labor themselves and sell at the local Amish-operated produce auction. They spray with organic fungicides. And, their numbers are far better than $850 per season per tunnel, almost 5× as high.

To be quite frank, it's wise to always estimate on the low side for profit margins when making calculations and estimating yields of crops. But do not let the terribly low profit forecasts of the conventional methods scare you away. Through organic production, doing most of the labor yourself (at least to start), engaging in strategic direct marketing, and making a name for your operation, you can easily smash these numbers and do much, much better.

To reevaluate a net profit forecast such as the UK tomato growing profit estimate mentioned above, first look at your own actual expenses. Do you already have a high tunnel? Are you/your family doing all the labor? OK, scratch off those two expenses. How much can you reliably and factually charge for organic tomatoes? Insert that number. Assume that, if you are a new grower, you'll get 20–50% of the estimated yield. If you are quite skilled, assume 75–90% of what is estimated. Remember, marketable organic yields could be lower, and in this particular example they are assuming you're growing extremely productive hybrid cultivars (such as the ones the universities would recommend). Their estimate is also assuming that you are spraying the crop with fungicides and insecticides. You *can* utilize organic approved spray products *and this is highly recommended, especially in humid climates.* If, instead of productive (non-GMO) hybrids, you're planning to grow lower-yielding open pollinated or heirloom cultivars, you can cut the yield estimate in half right now. Calculate these numbers together (yield estimate multiplied by price per pound, minus expenses) and you can get a close net profit estimate of what you could expect. Be conservative.

This is also an example of why cultivar selection is so important. If done strategically, it can drastically raise your yields and marketability. With the example of tomatoes, cultivar selection can double or triple your yields and also will determine your end product and marketability in general (e.g., cherry tomatoes or large slicer tomatoes). Hybrids almost always yield the most product by far.

To evaluate the gross profits from other crops, you'll need to know how much yield you can expect in your region and how much you can realistically sell it for. Factor in associated costs of nursery stock, labor, supplies, establishment costs (labor, equipment rental, irrigation lines, high tunnel, etc.). Subtract those costs from your gross profit estimate to get your net profit estimate. It will give you an idea of what's possible.

Understanding your bioregion and regional climate

Next, in order to be an effective grower, you'll have to understand the specifics of your geographic/climatic bioregion, *down to the particular microclimate of your site.* If you have not already, you'll need to observe, explore, and get to know it. Learn by association with other local/regional growers, *over the course of many seasons.*

No one can accurately say that in order to grow good blackberries, you toss them an ounce of 10-10-10 fertilizer every April for great results. Your soil type and micro climate can vary subtly yet crucially from someone even 20 miles away and you'll eventually learn how these small differences can make wide variations on the way species and exact cultivars perform. Temperature variations of 5–10°F (roughly 3–6°C) in winter from one nearby location to the next (even the top of a high hill to the base) can allow certain fruit species and cultivars to be either fruitful or killed outright in winter. Winter warm-ups like we have in the Mid-Atlantic region can spell doom for some cultivars and entire species, such as early-blooming apricots and almonds, and 'Pakistan' mulberry, and yet hardly affect others, such as the cold hardy 'Illinois Everbearing' mulberry and most other small fruits.

"Bioregion" here means the combination of a number of important factors, which include specific soil types, average and seasonal precipitation, local plant and animal life, winter and summer high and low temperatures, and local seasonal weather patterns (e.g., summer droughts, monsoons, mild year-round temps, extreme summers, etc.) Most regions have names. Our general region in Kentucky is the Bluegrass region. This region is noted for especially rich soils, epic pastures, lots of waterways, and hilly, limestone-based terrain. Broadly speaking we are in the Mid-Atlantic/Upper South with a Continental influence. This means we have fairly long, hot, muggy, usually moist summers. Here it's very lush and green and jungle-like; we generally enjoy rich clay-based loamy soils, fairly high precipitation (~40 inches or 100 cm per year spread evenly across the entire year), lots of wildlife, lots of humidity, and also very high insect pressure. The Continental influence makes our winters significantly colder (by about 5–10°F or approx. 3–6°C) than other nearby but more ocean-moderated places like Virginia or Maryland.

It's important to thoroughly understand the implications of your bioregion, especially because many of you will be farming in regions you did not grow up in. For example, most of California experiences a practically rain-free, warm to hot summer growing season with low humidity (basically March–October) and moist, very mild winters that hardly freeze. Conversely, in Kentucky, summers are usually fairly moist and very humid, and winters are moderately cold with lots of subfreezing, sometimes severe weather and drastic temperature swings. These (drastically different) regional factors exert major influence on your choices of viable species, adapted cultivars, farming techniques, timing of planting and harvest, etc. Your local ag extension office will help you understand and navigate these nuances, as will talking with other local growers and doing lots of in-depth research on your bioregion.

Hardiness zones

All U.S. regions are entirely mapped out according to their average *lowest temperatures*. In the USA this is called the USDA Hardiness Zone Map. There is a similar system in Canada and most other parts of the world. Each area is designated a number. These numbers correspond to the average "lowest possible" temperature for each zone, with a 10 degree difference between zones. For instance, in central Kentucky we are in zone 6, which means that the maximum extreme low temperature possible is –10°F (–23°C). That means that, on occasion, it has been recorded that the temperature in central KY can go down to –10°F. Yet, in extreme winters like we had in 2014, there were a few severe days of wind chill down to –20°F (–29°C), possibly colder. This is important to note. Usually, the extreme low temp is rarely reached but has been recorded.

Understand that your hardiness zone number is a *basic reference point* based only on winter low temperatures and nothing else, and utilize it as such. Other factors such as summer temperatures and conditions, winter chill hours, first and last frost dates, annual precipitation, etc., make a deeply substantial impact on plants and on agriculture in general, and are very important to consider.

For instance, parts of Washington state are in "zone 6," but those areas are more like highland deserts than the moist, humid, verdant pastures of "zone 6" Kentucky. Just because both areas are "zone 6" does not mean that what thrives in zone 6 KY will thrive in zone 6 WA and vice-versa, but many people out there assume this. Likewise, parts of

New York state are "zone 7," yet conditions are *very* different in "zone 7" New York from conditions in "zone 7" Alabama—namely, with respect to summer high temperatures, length of growing season and humidity, as well as local pests and diseases. Thus, there are serious limitations to blindly relying on Hardiness Zone maps to guide your decisions, yet they are an important basic starting point and tool to understanding your area. Research your region thoroughly and discuss it with local growers!

Reading and utilizing weather predictions and data

There are many online weather forecasting services out there these days. You might have a favorite one already. Personally, I utilize Weather Underground (wunderground.com). Many other good and free weather services are out there.

Their forecasts are shockingly accurate and created by bringing in data from thousands of localized mini weather stations across the USA and beyond. You can choose the station you want data from (or even get one installed on your farm!). Chances are there is one not too far away from you. Make sure the station you're drawing your data from is as close as possible—the website may automatically choose one for you that is not as local (and thus applicable) as others.

In general, I have found that our farm seems to get 1–3°F colder in winter than is usually predicted. This can make a substantial difference: 33°F is above freezing, 31°F is not!

Weather Underground, and likely others, also offer a History feature where you can see the temperatures for any day going back decades or longer. So, you can track your weather patterns, see general trends, look for climate change patterns, and see if your local climate is much different than it was in the past. Also, it has very clear and exact data on predicted precipitation, wind speed and direction, etc. As you progress in your fruit growing this data will be important, as it can help you make important decisions, such as when to open or close high tunnels, when to cover plants, when to plant, etc. Also, try to observe other local phenomena to hone your own intuitive sensitivity. For instance, I can hear distant train whistles when it's going to rain soon. The pressure change associated with incoming rain gives me mild headaches sometimes. Also, the lowest point on our land begins to be blanketed in frost hours before anywhere else, signaling a very cold night ahead.

You should also install accurate digital thermometers, especially ones that record the lowest and warmest temperatures detected. You can also get thermometers that send data to your phone via an app, which can be useful. That way you can track your local conditions and compare that to your forecasting info and see general trends on your land versus what your local weather station reports. Stick to one weather station and website, don't switch around.

Choosing cultivars

Eventually, in your planning process and research, you'll have narrowed down what species are viable in your area and best for your marketing plan. So, what regionally appropriate *cultivars of your chosen species* will grow best? This crucial question can literally *make or break* your operation or planting, of any size.

Cultivar selection is key to success because only specific cultivars of a species will be adapted and thus thrive in any given area. Other cultivars that are not adapted will completely or mostly fail due to vulnerabilities to local conditions such as winter low temps, frost dates, pests, diseases, etc. This is true for virtually every food plant in the world. Additionally, you need to ascertain whether your chosen cultivars are not only adapted but will also produce tasty and marketable fruit in your region.

Importance of cultivar selection

A "cultivar" is a cultivated strain (often erroneously called a "variety") of a plant. 'Golden Delicious' and 'Granny Smith' are two distinct cultivars of the species *Malus domestica* (apple). Cultivars are so *specific to region*, and *vary so widely*, that some will thrive and give bumper crops one place, and fail to yield in another. So very carefully research and decide what cultivars to plant in your region. You can start this research by contacting your local ag extension office for recommendations. Research what cultivars the local orchards, local fruit growers, and market farmers grow. Check online fruit growing forums and ask questions. Remember, if you are marketing fruit then backyard grower recommendations might be useful but may not be the best choice out there.

To illustrate the importance of this, I once read a book by an organic farmer in Los Angeles that left an impression on me. He said he made the mistake of buying fruit trees based on lovely nursery catalog descriptions such as "the highest quality," "excellent production," etc. He bought a few dozen of these peach trees and planted them out. In a couple of years, the trees matured and began to bear a crop. As the green, immature peaches swelled up with juice his anticipation grew. As the green peaches sweetened, ripened, and matured to the picking stage, and yet stayed *green peaches, never turning orange*, his anticipation turned to horror. He couldn't sell *green peaches*; no one would buy them, regardless of how sweet they tasted. Especially not in a super competitive market like in LA! In his region that *specific cultivar* of peach does not color up due to temperature or some other factor. He ended up selling them all as dried peaches to minimize losses and, disappointed, he tore the trees out. It was a waste. Don't let it happen to you. This is an example of a mostly regionally adapted cultivar but one that is not marketable (in this case due to coloration). Other cultivar issues could involve a tendency towards perishability, splitting, poor taste, etc.

Due to the rising numbers of new, mostly backyard, growers, cultivars are nowadays often designated by nurseries into variations such as: disease-resistant/good for home growers/commercial varieties. As a market farmer, you will often want to focus on the first two categories. Commercial cultivars are often productive but the quality may be lacking and they may be susceptible to disease, as most growers rely on synthetic fungicides, etc., to produce the crop. If a commercial cultivar is reliable *and* disease resistant in your area, then it may be good for your purposes.

But if the quality of the final product is not great, customer interest may be lackluster. Flavor is important and is one of the major draws of localized food production in general.

However, you do not need to focus on heirlooms or old varieties with "supreme flavor," etc. Flavor is important, but so is production, resiliency, and reliability. Many newer cultivars have it all, as many breeding programs (thankfully!) are starting to focus not only on production but also on developing better flavor and some amount of disease resistance. Some older cultivars may produce high quality fruit and be very resilient, however, hybrid cultivars are important and will continue to play a major role in the future of small farming and global agriculture. There is nothing wrong with growing hybrid cultivars or even hybrid species (such as modern strawberries and blueberries). These can often prove super productive and very high quality. Hybrid fruit cultivars are not GMO. A hybrid is, simply stated, the offspring of two open-pollinated cultivars or species. This creates a new cultivar (or species) that often will express high vigor and production. As of this writing, no GMO small fruits even exist,[3] except for a couple of tomato (and pepper) cultivars that you won't have access to anyway. So don't worry on that score.

Most of all, *regional adaption* is foundational. If a cultivar or species is not adapted to your region, then yields will be zero. You have to do your homework! The ag extension office can help with this. Also, never call a nursery in another state and ask them what will grow in your area. Nursery customer service employees are not nationwide fruit experts and they will usually give you very generic recommendations based on USDA Hardiness zones and this can rapidly steer you in a very wrong direction. You have to do the research yourself and do it well!

How many plants?

So, now we've come to this point in the planning process—narrowing down your bioregion, *check*; viable species, *check*; cultivars that will work, *check*.

Next, you'll need to calculate how many row feet you want to plant to a given fruit and how many plants that will take. Unlike what many online articles might suggest, it's surprisingly unreliable and difficult to ascertain how much profit can be made based on row feet, or square feet, of a crop. There are simply too many real-life variables. That includes your skill level as a grower, your soil type and health, growing season, regional climate, specific cultivar of plant, price you get for the product locally; seasonal climatic variations, if you're hiring labor out, etc. There are also seasons of average yields, bumper crops, and crop failures.

Again, visit a regional grower of a crop you want to grow and try to get as much info from them as possible, perhaps by posing as a curious customer. You could ask, "How many row feet or acres of (x) do you grow? What are your yields?" Then, with the data on price per pound and their yields, you can calculate an approximate yield/row/price. Or, if you start a small planting of, say, raspberries, you can plant a 100 ft. row and calculate using your own data. If you make a $1000 profit on the 100 ft. row, and could obviously sell much more than you produced, then it's reasonable to think that if the row footage doubled, and everything else stayed the same, the profit likely would double too. This has limits as the Industrial business model has historically shown us. Exponential growth creates a need for exponential labor, resources, and market, and that is impossible because we live in a finite system with limits. But, you're a market farmer (or backyard grower) so you're probably not interested in planting a hundred acres anyway.

If you are consulting with an ag extension office, they will assume you'll be using a tractor to mow or cultivate between rows of plants. So, row spacings (as well as fruit plants per acre calculations) will be based around fitting a tractor comfortably between rows. Your rows can usually be made much closer if you don't plan on using a tractor.

Generally, always run rows of plants North-South so the plants get maximum and even sun exposure on all sides. In some regions such as excessively dry or hot ones, this may differ, so check with local growers.

Don't plant rows or plants too close or they will shade each other out and lead to lower yields or other issues such as diseases. Around the height of the crop itself is a decent approximate measure of row distance if no information can be found. So, if raspberries are on a 5' tall trellis system, about 5–6' between rows should work. It's better to make rows a little wider apart than necessary than to crowd them together, as mentioned earlier. Crowding reduces sunlight access, airflow, work space (including harvest tubs, crouching farmers, wagons, etc.) and this creates frustration, causes plant diseases like fungal infections, increases populations of fruit flies like Spotted Wing Drosophila (herein referred to as SWD), and lowers yields. Likewise, excessive spacing is a waste of space, requires more maintenance in the form of mowing, results in less efficient workflow and lower yields, etc. Don't guess on this, research exact spacing and note what other growers are doing.

To calculate how many plants you need per row, *count how many row feet* you intend to plant and *divide it by the distance between plants*. For instance, if you're planting a 100' row of red raspberries and planning on 2' between plants, then divide 100 by 2 and you get 50 plants. Simple enough.

Pollination

This is an important topic that warrants your attention. Bees are a hot topic right now, due to their decline (many species, not just honeybees). Whereas in 2022 the honeybee fad is turning around and people are questioning the merits of the "exotic imported" honeybee, the fact is that honeybees are not only extremely important but originally native to North America.[4] However, controversy aside, all bees are important and matter. Especially when you are growing organic small fruits!

Honeybees do a great job pollinating many small fruits (blackberries, raspberries, etc.) However, you may not be able to raise honeybees, and honeybees are not very active in cool, moist weather when some small fruits bloom. Other bees and pollinating insects can do a great job ensuring your yields are high, if you allow them space to exist and some habitat. The best habitat is "shaggy", unmowed, and left alone areas with lots of native wildflowers. Allow the wildflowers to finish their lifecycle and do not tear down the dead plants until late spring (or not at all). They may contain pollinator or beneficial insect eggs! Even just a few "shaggy" square feet in the corner of your backyard can help if that's all you can spare.

Mason, sweat, carpenter, and bumble bees are all excellent pollinators, as are many types of wasps. We raise plenty of honeybees but also build simple little houses for the native bees. Some bumblebees nest in the ground. Don't kill wasps or carpenter bees. In this way, you can ensure good pollination of your crops. And, if you raise honeybees you'll have other highly valuable and marketable goods: honey, wax, propolis, nucleus colonies, etc.

A repurposed milk powder can, some hollow bamboo canes, a piece of hardware cloth and some old rope have created a sanctuary for wild bees, in minutes, for pennies. All the mud-capped tubes contain native wild bees/eggs. Kept under cover and protected from woodpeckers (via the hardware cloth screen), these houses help attract and keep pollinators around.

Summary

1. Do your research thoroughly. Decide what species you want to grow. Base your decisions on your local climate and region, your intended purpose, what the local market is doing, and what's in demand locally.
2. Consult with your local agricultural extension office to understand if your species will work in your area, and find out what cultivars they recommend. Talk with other local growers. Research university publications on the topic.
3. Understand how to estimate your potential profits and yields before moving forward.
4. Understand your local bioregion and climate. Explore your planting site carefully and understand your soil type, local weather patterns, and growing conditions. Choose cultivars based on this data and local recommendations.
5. Calculate how many plants you're going to need for the planting.
6. Facilitate, protect, and nurture pollinators, including wild bees and wasps by creating and protecting their habitat and nesting areas.

3

Sourcing Plants and Navigating the 21st Century Nursery Scene

Sourcing plants and plant material

Where and how do you get quality plants? This is a crucial question, no matter the size of your operation or the end goal. Let's find out.

As a professional nurseryman, I can tell you sourcing quality nursery plants at fair prices can be very challenging. You're dealing with highly variable living things and that always brings in lots of complex uncertainties. Due to seasonal weather a nursery's crop of plants might be subpar, and that happens to be the year you buy them. Or you may get scammed by a fraudulent online nursery or a shady local dealer. Or you buy plants that look good but internally harbor viruses that lower the yields. Be very, very careful when choosing nurseries to buy plants from. This chapter will help you navigate this challenge. Note that small plants in the nursery trade are often referred to as *starts*.

Consider shifting your thoughts from sourcing *plants* to sometimes sourcing *plant material*. *Plant material* is a nursery term that includes living parts of plants such as fresh roots or viable stem cuttings. You can save hundreds or thousands of dollars in establishment costs this way. But again, you have to have quality sources, know exactly how to properly handle the material, and be proficient at it.

Sources to avoid

Probably the worst place to acquire plant material is out of someone's backyard (either free or found online). Why? You have no idea what diseases, insects, or viruses they harbor that you will be bringing to your site, especially if soil comes with them. You have no idea as to the validity of the cultivar name. *Don't even think about it*, even if it's your best friend's raspberry patch. You're running a business (market farming) and have to be serious and careful about it. A lot of work, time, and capital will go into this. Don't get off to a terrible start by starting with random sources and/or questionable stock that may lower yields or fail. Even for a backyard planting, start with fresh, high quality stock only.

If you have some basic horticultural skills and time, you might consider saving a lot of money by starting plants via *root cuttings, stem cuttings, or virus-indexed tissue-cultured starts.* These can establish strongly and grow rapidly with proper care and some attention. How to start and grow all these is beyond the scope of this book, but some specifics, such as growing blackberries via root cuttings, are mentioned later. In general, primarily only blackberries and raspberries are grown from root cuttings in the field. Tissue cultured starts can be used when growing figs and strawberries, and sometimes are available for other fruits also. They are usually virus-indexed and clean of all diseases. Tissue cultured starts are generally very small, maybe only a couple of inches tall, but grow surprisingly fast in good conditions, and also are often very affordable.

Important to note is that these days, many superior cultivars of plants are "patented plant varieties" (PVP), and may be under US or other patent protection. That means you are not legally allowed to propagate them via cuttings or roots without a license or permission from the patent owner. You can easily check for plant patents, as well as patent status, on google, by typing in the plant cultivar name and "patent" after it. Plant patents generally last for 20 years, after which time the plants become "public domain" and unrestricted to freely propagate. Free the plants!

Organic or conventional?

Discussions about nursery stock often come to this pertinent question. It actually doesn't make a big difference, in the long term, for your planting, as long as the plants or material you source are healthy, fresh, and well grown in general. Plants don't "care" whether the nutrients they receive are organic or not. They'll grow practically the same, although the quality of the end product is usually or, at least, often superior with organics. The real concern is the environmental and social impact of the nursery trade and other forms of industrial agriculture. That is beyond the scope of this book, but I am sure most of you are well aware of the devastating effects of conventional, non-organic, industrial agriculture on the planet's resources, land, water, air, and people.

Whenever possible, obtain high quality, organic plants as available and practical. Make sure the quality is truly there: I've seen plenty of pitiful organic nursery plants and plenty of amazing conventional ones, and vice-versa. Be aware too, as of this writing organic nursery stock availability is extremely limited, often not available for many species, and usually only available from retail nurseries at very high prices.

Our nursery business, Peaceful Heritage Nursery, is USDA Certified Organic, so you know where we stand. However, non-GMO, yet conventionally-grown plants and starts can and do provide excellent results when well grown under organic conditions. Don't settle for incorrect or non-specific fruit cultivars or varieties just so you can stay with sourcing organic plants. This will be a big mistake. If necessary in order to source the exact cultivars needed, buy conventional starts. Then, just grow them under organic conditions and it will be all good. The plants you source are just tiny starts anyway, and the mature plant and what they produce will be organic and wholesome.

If applicable, check with your organic certifying agency on current rules around sourcing non-organic plants and starts. Currently they are acceptable due to the vast lack of availability of organic plants. You just have to prove you put in the effort to source organic nursery stock and could not locate what you needed, usually by citing 3 legitimate sources you checked.

Navigating the nursery scene

Finding high quality nurseries can be a challenge. Unfortunately, many online nurseries are low quality and/or overpriced. Many local nurseries don't sell bare-root plants or the cultivars/species you're looking for. Where to go?

I can tell you a few places to virtually *never* buy nursery plants from. These include Etsy, eBay, Alibaba, Amazon and any other online boutique type outlets or most auction sites. These places are scammer heaven for a number of reasons. Anonymity, lack of quality control, and lack of professional skill abound in these online arenas. In other words, expect lots of "backyard nurseries" selling low quality stuff. Expert growers and professionals virtually never buy (and rarely sell) anything related to plants or seeds through these websites.

Just looking at the listings on these sites should clue you in that a majority are offering poor quality or simply fake products: blue strawberries, rainbow colored tomatoes, giant fruits the size of your head, etc. It's easy to scam on these sites because this is where newbies and clueless people shop for plants and seeds. Even if they're not fake, the plants could harbor viruses or be low quality. Avoid.

The many fraudulent online nurseries are another terrible source to shop. Despite a professional looking website, they often sell distressed or dying plants, will ship you the wrong species or, at best, the incorrect cultivar or subpar specimens. Customer service is non-existent or unhelpful. Refunds are not an option. If an online nursery features overly-hyped descriptions, off-putting or severely outdated photos, or obviously fake photos, two-for-one deals, strange sounding cultivars or exotic plants no other nursery sells, or just seems very generic or shady, avoid them. These are all clues you're most likely viewing the website of an unscrupulous or low-quality nursery. Check lots of online reviews and forums before making a purchase.[1] If a nursery has low ratings and terrible reviews overall, strictly avoid them. Every business gets an outrageous, negative review from time to time, but if the ratings are obviously in the majority or abundantly negative (like a 1–3.5 star rating out of 5), stay far away. You could waste a lot of time and money, and fund an unscrupulous business. Always review the Terms and Conditions of a nursery before making a sizeable purchase. If there are none available to read, or they are unreasonable, *go somewhere else.*

I can also tell you a clue from my professional experience: if a nursery sells *both* fruit plants and ornamental plants, that is a big red flag. You want a nursery that specializes exclusively, or let's say 95%, in fruit. That should be the focus of their business and what they do. There might be a few exceptions but, generally, selling both fruit and ornamental plants is a sign you're dealing with a business that is simply selling generic plants, is not expert in fruit growing, and likely deals in lower quality or overly expensive product. You want to buy *only* from professional fruit nurseries, preferably nurseries that propagate all or many of their own plants. These are the people who know fruit and sell quality plants!

Other very unsuitable places to source from are *big box stores* or nurseries that mostly sell ornamentals and *happen* to have some fruits also. The employees at these stores usually know absolutely nothing about fruit growing and simply sell whatever they receive. The plants are potted in chemically fertilized peat moss, are distressed and also very expensive. You can't afford to pay $15–20 per blackberry bush when you want to plant 25, 50, 100, or more. Through wholesale purchasing you can obtain healthy, virus-free, bare-root blackberries for half that or less, sometimes as cheap as a few dollars per plant. These will also grow and establish much better than the potted, chemically grown plants.

Quality nurseries

So, where to go? Check wholesale nurseries (with great reviews) if you're buying a lot. Large nurseries that have been in business a long time and that specifically serve commercial growers are usually very reliable, otherwise the commercial growers would not utilize them. However, commercial wholesale nurseries often lack the cultivars you may want and they can sell out rapidly. Many smaller nurseries can be great choices, but check their pricing, online reviews, and ask questions. Are the plants certified virus-free? Are they containerized, or field grown and shipped bare-root? What is the guarantee on the plants? When do they ship to your region? Are there bulk discounts? How long have they been in business? Are they regional to your general area? Do they seem very knowledgeable and trustworthy or generic and shady? These things all make a difference.

What to source are "certified virus-free plants" that are *field-grown* (preferably in similar conditions to where you will be planting them, although this is not always possible) and *shipped bare-root*. Certified virus-free plants are not always available, and usually only are available for the newest cultivars. Likewise, sometimes potted stock can be high quality (yet much more expensive per plant). Why bare-root? In general, very few berry or small fruit farmers utilize potted plants, with the exception of blueberry growers, as it's far too expensive to ship these in, unnecessary, and drastically more labor intensive to plant potted plants over bare-root. Bare-root plants are much lighter in weight and thus cheaper and less fuel-intensive to ship, with less packaging materials needed. They arrive in a clean, hygienic state less likely to harbor invasive species, diseases, and weeds. They have been field grown in natural soil, thus the root systems are much larger and more robust, and they will transplant better. They likely have undergone much less stress than potted plants in artificial conditions. And, they can be propagated utilizing much less land (and water), as field growing will always be more efficient per square foot than growing plants in containers. So, with bare-root you get a better plant produced and shipped in a more environmentally responsible way. As long as the nursery handles the stock properly once it is dug and ships it carefully, it is a far better product than a potted plant.

Important to note also are the realities of purchasing life forms. Too often I have customers purchasing plants online from us like they purchase shoes or other factory-made goods. They place the order and expect it to arrive in a few days or a week regardless of time or season, as though it was simply sitting on a shelf and had to just be packaged and shipped. When you are buying from a nursery, especially field-grown, bare-root, living plants, they will arrive when they're ready and in season. A nursery in the south will have plants ready to ship in February, and may do so in February or March. This could pose a big problem for you if you're a grower in Vermont, where it's not going to be suitable to plant until May. If you live in the North, avoid buying from a nursery in the South, if possible, unless they are holding plants in a dormant state in refrigerated storage and can get them to you in good shape months later.[2] Likewise, if you're a grower in Florida, avoid buying from northern nurseries, if possible. You will be ready to plant in February, but they may not ship until April. This could pose a problem for you, because by April it's already getting quite hot.

If you're a Northern grower, inquire if bare-root plants are stored under refrigeration. If you're in the South, inquire into the earliest possible shipping dates the nursery can handle and make sure things line up before buying. Just make sure before making a purchase that the plants will be getting to you within the optimal timing for planting in your region and not months early or late.

Again, always read over the nursery's Terms and Conditions before making a purchase. If there are none available for viewing, or no refund/replacement policy in general, this is a big red flag.

Always try your best to buy regionally. You can also save on shipping costs and minimize potential shipping related issues this way, as well as lower the carbon footprint of the exchange. You may be able to pick an order up in person, if possible (always contact the nursery to ask them before purchasing). See Chapter 6: Planting Successfully for full information on inspecting your nursery order once received, and how to handle it properly and plant correctly.

Summary

Make sure your plant choices, both species and cultivars, are thoroughly researched and check nursery sources *very carefully*. Only buy plants from a reputable company you feel very confident in—whether organic or not. Thoroughly check reviews and consult with other growers. Buy wholesale, dormant, bare root plants whenever possible, or consider utilizing plant material such as micro (tissue-cultured) starts, roots, and cuttings. Bare-root plants are generally far superior to containerized, and have many important benefits and advantages for you as the grower, as well as for the nursery.

4

Creating Beds

Creating space for small fruits to thrive

Growing in beds and rows

Creating spaces where your fruit plants will grow well is not overly difficult if you know how to do it right. There are basically two options for in-ground growing: *beds* or *rows*. Cleared, prepared spaces where the soil has been broken up/tilled, amended, and worked to a fine tilth are called *beds*. Rows, for our purposes here, are less intensively worked or may not be worked at all, and are simply rows of larger shrubs or small trees such as mulberry, figs, gooseberry, blackberry, or grapes, usually planted in individually dug holes in a row configuration. Rows for larger plants don't require extensive tilling or soil prep, but do benefit from cover cropping and organic fertilizers, as will be explained later.

Creating beds and rows

When you look at your site, you'll probably see lots of grass, unless you are on an arid or semi-arid site. If there's grass, it has to be removed or killed off. First, we'll look into mechanical means, then easy no-till options.

Grass is intensely competitive with your crops for water and nutrients. If you're planting *rows* for larger plants, you can accomplish grass removal quickly with a metal pick (or spade), shallowly removing the sod in chunks to create a cleared spot. Dig a hole in the cleared spot and

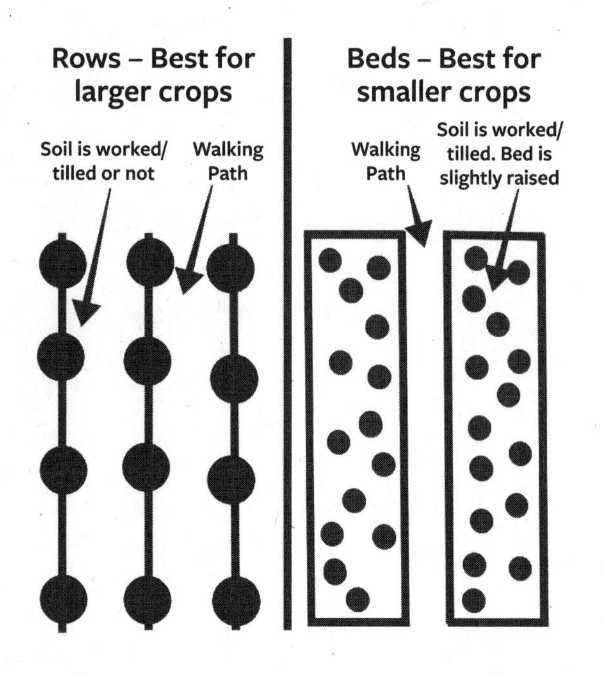

Crop	Bed/Row?	Planting Distance
Blackberry	Row	Erect 4–6', Semi erect 6–8'
Raspberry	Bed	4'
Blueberry	Bed or Row	Lowbush 2–3' Highbush 4–6'
Gooseberry	Bed or Row	3–4'
Currant	Bed	4'
Fig (indoor)	Bed or Row	4–8'
Fig (field)	Row	18–20'
Muscadine	Row	18–20'
Tomato	Bed	Field 5–6' Tunnel 10"–3'
Strawberry	Bed only	1'
Juneberry	Row	4–6'
Mulberry	Row	Bush 4–8' Trees 15–20'

then plant. Holes dug about 18" (46 cm) wide and 18" deep are sufficient for larger plants like those just mentioned. You can turn the sod upside down and put it underneath a thick layer of mulch to rot, instead of removing it (and the rich topsoil often attached).

If you are planting small, closely spaced plants such as red raspberries, strawberries, tomatoes, or other smaller plants, you need *beds* so the roots can easily expand into loose, fertile soil and, for some species, so they can then easily make suckers that will fruit in the future. Plants that have considerable space between them, such as blackberries or grapes, don't really benefit from beds and this may be unnecessary and increase weed pressure. Explore what other regional growers are doing.

Mechanical tillage

When making beds, the best width for the bed is 30 inches (76 cm). That makes it easy to reach both sides of the bed without straining your back when weeding, harvesting, etc. Creating a cleared bed can be done a number of ways, depending on your resources and timeframe.

First, there is the *mechanical approach*. If the site is covered in grass, you can rent a sod cutter for the day from a local machine rental shop for less than a hundred dollars. Run the sod cutter in straight lines across the grass or lawn, creating 2' (61 cm) strips of cut sod. It can now be peeled back and removed. If the soil is soft and moist underneath, you can use a tractor with a rotary plow to till the soil, or a hefty roto-tiller usually works as well. Tilling hard, dry ground is dangerous and may make the tiller jump on you, so avoid this. You can deep soak the future beds with a hose or sprinklers and come back and try again in 2–3 days, allowing the soil to drain a bit. Never till wet muddy soil, as it will damage the soil structure.

Credit: Earthtools.net

Making beds with a rotary plow attachment on BCS walk-behind tractor.

Make several passes over the entire area to make sure the soil is heavily tilled and broken up. You should not see large, hard clods of soil. If you do, make another pass or two. Till the beds, having the soil "shoot" into the bed, to create a ridge. Till the other side, shooting the soil into a ridge going the opposite direction. Now smooth flat and straight using a garden rake or with a bed-making tractor attachment. Growing beds should run north-south.

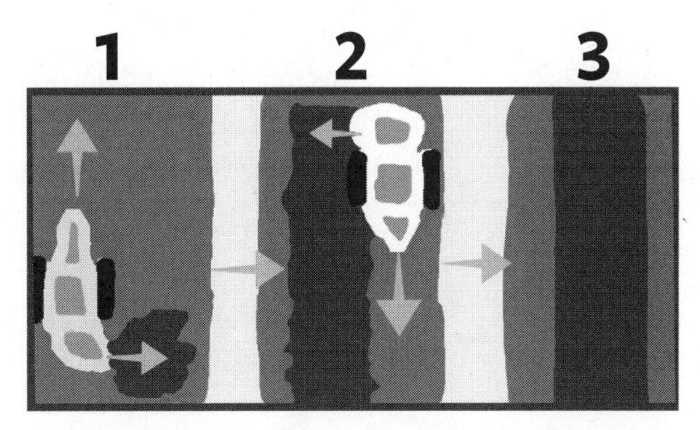

(1) Plot has been tilled and bed creation is underway by making another pass, plowing soil into a ridge. (2) Making a pass the opposite direction, the rotary plows shoots more soil into the same ridge. A rough bed is now made. (3) The final step is raking it smooth and flat.

If you want something larger to work beds with, a borrowed, rented, or hired tractor with proper implements makes tilling and creating beds fast and easy. Beyond a tiller, the needed implements for making beds could be: disc hillers, box blades, rotary plows, or bed shapers. Some of these are available with walk-behind tractors too. Just make sure the adjustable blades are set at about 30" bed width or whatever width you desire. Check the beds after making a few feet of bed to make sure the width is correct, and adjust as needed.

Mechanical tilling works fine for relatively larger plantings or when you need the land prepared very rapidly, because you can have beds ready within a few hours. However, tilling and plowing have repercussions on the soil structure and soil life that are detrimental (although often rapidly repairable). It has pros and cons, the "pros" being it's fast, easy, and effective. The "cons" are that it will affect the soil structure negatively. However, compost and avoiding further walking on the beds and compacting them will allow the soil to rapidly heal. Your fruiting plants are long term crops and the soil will not need to be worked again until you replace the crops or phase them out. At that point a broad fork can usually work the beds well and easily.

After tillage, you need to rake out any rhizomes, stones, debris, trash, etc. A shallow-toothed, wide, aluminum garden rake works best for this. If rhizomatous, invasive grasses are present, and you will want to remove as much of their white rhizomes as possible and dispose of them off site, or pile them up in the sun to dry for several weeks, or burn them.

Once the plot is cleared of sod and weeds, broadforking the beds breaks up any compacted subsoil. A rigid garden fork (*not* a pitchfork) can also be used (broadfork is easier). Broadforking is actually fairly easy and pleasant, although it looks difficult. Don't torque the fork too hard or the handles may break off. A 25' (7.6 meter) bed can be broadforked in 5–10 minutes easily.

Most fruit crops can be grown on prepared, flat beds just *slightly* raised above ground level by a few inches. If your site has inferior drainage you may want to raise the beds higher, up to 4–6" above the ground. Red raspberries also benefit from this. It improves soil drainage, which is critical to this crop.

If you are needing very deep, tall raised beds (usually 6–8" high) you can use a tractor-mounted disc bedder or raised bed maker to shape the beds rapidly. Or, you can rake them into shape with a rigid *garden rake*.

A rotary plow, like the one we use on the walk-behind BCS tractor, can also create high raised beds rapidly if operated correctly. You just have to make several passes back and forth and make sure it is "shooting out" the soil *into* the bed and not outside of the bed. A rotary plow is easier and safer than a tiller, and disturbs soil structure less than a tiller does.

Rake flat and smooth using the flat top part of the garden rake. Tying a string line to stakes and putting the stakes at both ends of one side of the bed you are raking can help make the beds nice and straight.

In arid sites you can *do the opposite*, digging out the beds a couple of inches (5 cm) deep so the beds are basically inverted and thus capture additional rainwater and have reduced moisture evaporation. If irrigating heavily this may not be needed. Check with local growers.

Amending the soil

After tilling, broadforking, and raking the soil flat and smooth, you can add compost and soil amendments such as granulated manures, rock powders, or vegan plant-based fertilizers. Avoid initially adding much high nitrogen fertilizer when planting perennial crops like fruits. You don't want to encourage overly rapid top growth at the expense of roots, and this can happen if they're overdosed with nitrogen. 15–20 US gallons (approx. 55–75 liters or 3–4 large buckets) of compost is adequate to amend a 25-foot bed for establishing new plants. The following year organic fertilizers should be applied throughout the growing season. If your soil is poor or has very low fertility, initially add compost and balanced organic fertilizers as will be explained later.

No-till options

There are also *no-till methods of bed preparation*. The best method for small scale is the *soil-tarping method*. Basically, you place an opaque tarp on your site to block out the light on the ground, smothering and killing all grasses and, with enough time, most perennial and annual weeds. On heavily grassed and weedy land, about 60–90 days of tarping will render

Three huge silage tarps placed in winter to prepare this no-till bed for spring on our farm. Note logs holding it in place.

the site completely clear, or 30–45 days in hot sunny weather, sometimes faster. When the tarp is removed, the site should look like completely bare, moist soil. Tarping can be done any time of year and if done overwinter does a great job at killing grass and preparing a site for spring planting.

For tarping, it's best to utilize *silage tarps*, which are used in the fermentation of hay into silage. These are fairly thick, durable, opaque, plastic sheets. Use 5 or 6 mil. Usually, one side is white and the other black. If the black side is facing the sun, it will warm the soil and also the heat will kill the grass faster. In hot, sunny summer conditions a black tarp may render a site clear in 30–40 days, possibly even faster in hot sunny climates. Make sure that the tarp is secured down all along it's perimeter with weights such as old boards, logs, sand bags, cement blocks, chunks of old cement or bricks, T-posts, sod staples, etc. It can and will rapidly blow away if not secured! Black tarps also heat up any rainwater that has collected and pooled, thus usually preventing mosquitoes from breeding in it. Any opaque tarp will work, including thick, higher grade hardware store tarps, although these are pricey and degrade into microplastic within 3–4 seasons of use. So best to avoid those if possible. For very small plantings, flattened cardboard also works. Some materials may not be allowed in Certified Organic operations, so inquire with your local inspection agency if going for certification.

Clear plastic sheets such as old greenhouse plastic also works well. Weigh down the sheets tight along the entire perimeter. When done like this on moist ground, it also solarizes the soil underneath. Solarization creates a greenhouse effect and steam-bakes the soil a few inches deep, thus killing all disease pathogens, nematodes, and most weed seeds. Solarization takes at least 6–7 days of translucent tarping during intense daily sun.

The great advantages of soil tarping are that it works while you do other things, uses no chemical herbicides, and takes minutes to set up, even on large areas of thousands of square feet. Sileage tarps are highly reusable and last for years. In addition, tarping also encourages earthworms and increases the soil organic matter through the breakdown of the smothered plants. When removed, the soil is usually moist and often more friable underneath. It is the main method we now utilize. You can also flame weed down any little weeds that come up a few days after the tarp is removed, or once beds are created.

When setting it up, you can either tarp off an entire large area to have a "blank slate," or cut the silage tarps according to your bed width (30 inches is recommended), secure them down where you want the beds to be, and leave the grass in between the beds uncovered so it does not get killed by the tarp cover. That way you are only killing off the grass in the future beds and leaving sod strips as walking paths in between the beds, which therefore do not require any reseeding, and can keep out noxious weeds from establishing. Keep the beds covered until all vegetation is killed back (30–90 days) then remove the tarps, and work the beds as described.

After tarping, you can use a tractor/BCS with tiller to make the beds as desired (described in the last section). Or, for smaller plots, you can broadfork the future beds and then shape them with a hoe and garden rake (thus avoiding any tillage whatsoever). Tarping leaves the soil easy to work. If tarping to plant in rows, you can now plant your shrubs/vines/trees in rows and then sow low-maintenance plants like ladino clover, comfrey, or manageable grasses to revegetate. Or, cover in thick organic mulch.

Spacing

When spacing, about 6–8 feet (1.8–2.4 m) between the beds or rows allows proper space when growing taller plants like raspberries and blackberries. Often recommended in university literature you will see 12' (3.6 m) between taller crops such as these being advocated, but unless you're mowing with a large tractor, that is way too much space in my experience.

You do not want the beds/rows overly close together or the crops will shade each other out and this is very detrimental to yields and crop health. All parts of the plants need to receive direct sun (except the very bottom).

Bed spacing differs based on what you're growing. Low growing strawberries can be planted in beds that are spaced only 18–24" apart (approx. 45–60 cm) or possibly even closer. Large plants like raspberries may need 5–6' (1.5–1.8 m) between beds. Be very specific with bed spacing and thus the width of the walkways.

That being said, having the beds or rows too far apart wastes space and creates more mowing; too close reduces health and yields. Plan accordingly and carefully per species to find the balance between too much and too little space. Consider also your method of grass/weed control in between beds/rows—spacing needs to be adequate to fit any equipment.

Managing walkways

To manage weeds in walking paths in between your beds, you can either seed the paths to a non-rhizomatous ground cover grass like fescue, orchard, or bluegrass (or other regionally appropriate grasses) and keep it mowed short. Or, you can utilize a low-growing, traffic-tolerating ground cover plant like ladino (white) clover. Grass does a better job long-term at outcompeting weeds and handling foot traffic than most ground covers and lasts indefinitely. Keep in mind, string trimming or mowing grass is time and labor consuming and can toss clippings onto low growing fruits like strawberries.

You might also choose to cover the walking paths with landscape fabric or other materials, so it stays maintenance-free and weed-free year-round. You can also keep them bare, and manage the weeds with a sharp stirrup or hoop hoe fairly easily, if hoed when the weeds are very small. This actually works quite well, especially for very closely spaced beds such as those for strawberries. This would not work for large, wide walking paths. Organic mulch/flattened cardboard is also an option.

Stale bedding

One strategy to consider is to allow the prepared beds to sit for 10–14 days during warmer weather, before planting. Soon many tiny weeds will likely emerge. Flame weed them down (or stirrup hoe them), *then* plant. This is a good option for heavily weed infested sites and can save a tremendous amount of future weeding labor. This is called "stale bedding."

Planting the beds

Plant the beds as soon as they are prepared, or stale bedded and flame weeded, so weeds cannot establish. Carefully time this to the arrival of your nursery stock or planting time in your area. If your shipment is arriving the week of March 15th, then about 3 months before that date (Dec 15th), tarp your site. A few weeks before the stock is to arrive prepare your beds properly (including stale bedding if desired) and then plant as soon as possible. See Chapter 6: Planting Successfully.

Planting in rows

Larger plants and shrubs (muscadines, blackberries, gooseberries, etc.) can be planted in rows and do not require bed creation. These can simply be planted in a spot that has had the sod removed and a nice large hole dug (usually 18" wide and 12–18" (30–45 cm) deep suffices). Sod can be removed with a pick or mattock, removing an area of sod slightly wider than the size of the hole. Planting holes should be spaced apart adequately per species. Rows can be kept straight with string lines, running north-south. Alternatively, you can till entire rows (or tarp for no-till) in order to loosen soil and eliminate weeds. It's almost never necessary to do this. Weed management is covered in Chapter 7: Organic Weed Management.

Cover cropping

You might consider cover cropping for a few months or even a full season or longer before preparing your site, especially if you're not in a rush. Or, if you are expanding onto multiple sites (or phasing out an old bed of strawberries, etc.) you can prepare the new one by cover cropping it. Cover cropping deeply improves soil tilth, increases soil organic matter and, thus, the moisture retentive capacity of the soil. It also increases resources for the flourishing of soil micro-life, increases available nutrients (mostly nitrogen) and enhances bioavailability of minerals like phosphorous. Nothing improves soil quality faster, easier, deeper, or more sustainably than well-executed cover cropping. It's easier to do than you might think.

If the site has been tarped/tilled and is clear of vegetation, cover cropping is often as easy as thickly broadcasting the cover crop seed onto the bare soil. A heavy rain or watering and, if available, applying some very light straw mulch gets it off to a strong start.

Optimal cover crop species and timing varies by region, so check with your local agricultural extension office or other professional growers. Grass and legume mixtures are common and often the best option for most regions (e.g., annual rye with hairy vetch or oats with crimson clover). This very potent "tag teaming" performs double duty: grasses protect and break up tough soil and hardpans, increasing tilth through adding thousands of pounds per acre of organic matter[1] while the legumes stimulate soil life, feed pollinators, and add nitrogen and organic matter. Buckwheat is a great and easy to grow short term cover crop that improves soil tilth, rapidly protects bare soil, and matures in only about 40–50 days. Legume and buckwheat flowers also serve as highly nutritional forage for honeybees.

The cover crop should be terminated (killed) at flowering via mower/sicklebar (preferable) or scythe (very small scale only), or flattened/crimped and then, on smaller plots, it can be immediately tarped for 3–4 weeks in spring/summer. After removing the tarp, the cover crop will be partially decomposed except for the dried residues on top that will look like straw. This thick straw layer can be raked and used as mulch in the beds/rows. Very nice! Make sure the cover crop is 100% terminated before planting! If it is not and it starts to regrow it can compete with your crop and become a real issue. Once the tarp is removed, you can start making the beds. Cover cropping can also be done in between rows, as is done in California wine country, utilizing mustard. Likewise, make sure the cover crop *is* actually terminated. If it is not completely killed, it can regrow and become a weed problem of it's own. This is especially true of many grass cover crops.

The main mistake I have seen small growers make with cover cropping is not terminating it early enough, so that it gets fibrous and very difficult to remove (or goes to seed and becomes a weed). Buckwheat can easily become a weed if not terminated shortly after blooming (about 30–35 days from sowing). So, make sure to terminate at blooming all cover crops!

5

Managing Soil Fertility Organically

Where will your soil fertility come from? You'll need *organic nutrition for your crops beyond what the local soil will provide.* Let's focus on practical ways to effectively get your crops the nutrition they need to perform their best.

There's no question that organic inputs, even industrial ones (such as chicken manure from a conventional farm or soy powder from industrial soybean farms) are overall more nourishing to the soil and micro life, healthier for the environment where they are being applied, and safer for the farmer. There's absolutely no comparison to being back sprayed in the wind with chrysanthemum extract or neem oil versus being back sprayed with 2,4-D. Exposure to the former is unpleasant, the latter causes severe, painful long-term illness and possibly cancer.[1] I think we're all clear on this point. In fact, I once met a homesteader who worked in a conventional nursery and this very thing happened to him. He was sick in bed for a year. It took major detoxing and lifestyle shifts to get him back on his feet.

Practical considerations

You'll need to eventually base your fertility program on what you can obtain locally or regionally. If you are vegan and/or have an aversion to animal-based products (for instance, on our farm we don't use any slaughterhouse products like fish, bones, feathers, hoof, or blood) then look into *plant-based vegan* fertilizers. These are often more expensive but, in our experience, do an amazing job.

Don't overdo fertilization of your crops and, generally, you should stop applying all fertilizers by June or July so as not to stimulate late-season flushes of soft growth that will fail to harden off and lignify into wood before freezing weather. This will depend on your location, so check with local growers.

Soluble and insoluble organic fertilizers

Organic fertilizers are sourced from various natural products, byproducts, and natural resources, both renewable and non-renewable. The first major distinction is between *soluble* and *insoluble* fertilizers. Soluble simply means it rapidly dissolves in water, becomes a solution, and thus can be sprayed or injected into irrigation water and used to "fertigate" (fertilize through your irrigation system). Soluble fertilizers will not clog your irrigation sprinklers, drip lines, etc., although filtration on your irrigation system may be useful. Soluble fertilizers are rapidly and highly effective, uniform, and efficient, especially for crops grown in high tunnels or containers, but also in the field. You'll often see results within a few day's time.

Fertigation injectors are available and can turn an existing irrigation setup into a fertigation system for a couple of hundred dollars and a few hours labor. Fertigation is increasing in popularity as a technique and is easier and cheaper to implement than you might think. Fertigation systems are easy to install, save a lot of labor, and deliver consistent, uniform results. See the Resources section for recommendations.

Insoluble fertilizers are simply powders, meals, or granulates that will *not* mix with water or become a flowable solution. Avoid using these in fertigation systems or they will clog your system and may damage it. Insoluble fertilizers are slower release and for direct application or mixing into potting soil, growing beds, fields, etc. Organic-based insoluble fertilizers like soybean meal, feather meal etc., once applied, generally begin to break down within 1–2 weeks and start feeding plants at that point, and continue providing nutrition for a month or so, maybe less in sandy soils. Rock or mineral based products (rock phosphate, greensand, etc.) take longer to begin feeding plants (minerals only, no nitrogen) but last for years as they very slowly decompose.

Animal-product based fertilizers

The next distinction in organic fertilizers is between those sourced from animal products or derived from vegan sources (plant or mineral-based).

Typical animal-based products/byproducts include liquid fish (emulsion), as well as blood, hoof, feather, crab, fish, and bone "meals." Meal simply means dried and ground into powder. These examples are mostly slaughterhouse or industrial fishing by-products and often have heavy carbon and karmic footprints. Morally, I cannot recommend using those. Some animal products are not slaughterhouse byproducts, and include

	Vegan Fertilizers	Animal-Based
Pros	*Ahimsa*	Very effective
	Fast breakdown	Strongly stimulates soil life
	Very effective	Can be higher in organic matter (manures)
	Hygenic and no smell	Easier to obtain
	Doesn't attract animals	Often longer-lasting
	Easy to apply	Can be produced on-farm
	Possibly lower carbon footprint	
Cons	Can be pricey and challenging to source	Many are slaughterhouse by-products
	May require more applications	Can attract rodents and pests
	Organic sourced may not be available	Many rules and stipilations with USDA Certification
	Often local/regionally unavailable	Can be smelly and unhygenic

eggshells, local animal manures, or your own farm/livestock byproducts. Chicken and turkey manures as well as cow and horse manures make excellent, well balanced organic fertilizers.

Poultry manures in particular (turkey, chicken, seagull) are very potent, complete fertilizers and give outstanding results simply used alone. They are high in calcium and nitrogen, and really get everything growing fast and strong. Seagull and bat guano are especially high in phosphorous, and are sometimes lacking other nutrients, so go for ones that include NPK (nitrogen, phosphorous and potassium) and not just P.

Cow and horse manures are much lower in nitrogen than poultry manures and slower acting, but increase organic matter levels and soil-life populations much more than chicken and turkey manure. When used in conjunction with bird-based manures, even better results occur.

Liquid fish emulsion is a very fast acting stimulant to soil life and high in nitrogen and trace minerals. It is a general panacea for many nutrient deficiencies, including nitrogen. Weakling or slow-to-grow plants can be stimulated rapidly with an application or two of diluted fish emulsion. We previously used fish emulsion but have fully transitioned to using fermented vegan corn and soy water soluble powder fertilizers with added kelp extract. We prefer its gentler, yet very effective, action and its lack of foul fishy smell, and have seen amazing results using it.

Vegan plant-based fertilizers

Typical *plant-based vegan fertilizers* include fermented soy and corn-based products, molasses, alfalfa meal, kelp, composted crop residues, and seed meals (cotton, neem, etc.). These are a great option and the quality and availability continue to improve. We are gradually phasing out animal products in favor of vegan options, with impressive results.

Mineral fertilizers

Mineral fertilizers, or rather soil amendments (which are also vegan) include greensand, ground limestone, Tennessee Brown Rock, basic slag, rock phosphate, azomite, epsom salts, etc. Mineral fertilizers are carbon-free and therefore generally only provide trace minerals or potassium (K)/ phosphorous (P) but not nitrogen (N). Nitrogen is essential for green leafy growth and must be supplied also. Minerals are best used for long-term replenishment of minerals on damaged, exhausted, or previously farmed soils; new sites; or those depleted of (or naturally low in) crucial minerals like magnesium, phosphorous, etc. Kelp, a quick-releasing biological trace mineral supplement, is useful for *increasing the sugar/mineral content of the crop* and therefore the *brix* (dissolved sugar) of the fruit. Rock minerals can affect the pH of the soil, especially limestone (alkalizing), and can sometimes provide too many minerals if used excessively, so be selective and careful when adding them, as you cannot remove them. Always do so for a specific reason, often based on soil tests. Some mineral amendments can contain harmful levels of heavy metals, so check your sources very carefully.

Compost and manures

Local, well-made, and *finished compost* is also an option. Finished compost should have *no visual trace* of any parent materials. (Obviously visible sawdust, animal manures, wood chips, etc., means the product is NOT finished compost and is only partially decomposed). Real, finished compost should smell sweet or earthy like forest soil and should never smell sour or foul. It should look dark and crumbly, and *you should not be able to tell what it came from*. Anything other than that is *not* finished (actual) compost and will be a waste of money per cubic yard (if buying it), and it *may lead to issues with your organic certification or plant health*. Even if it's been sitting piled for a year or longer, that doesn't make it finished compost. It should also not contain noxious weed seeds. (Be careful of overgrown compost piles surrounded by weeds!)

If non-decomposed sawdust or wood chips are present they can "tie up" nitrogen in the soil until soil bacteria degrade it. This can cause nitrogen deficiency in the crop, visibly stunted growth, yellow leaves, etc. Check products, sources, and the composting processes carefully. Go for professionally made compost if available. Inspect before you buy.

Sometimes local ranches and farms will provide tractor scoops of manure for free or for a small fee, and this can be a good source of fertility, but it is labor intensive to apply and must be composted first. Compost has to be used in prolific quantities in order to suffice as a general fertilizer if not using other more concentrated animal or vegan fertilizers. Those large quantities can be difficult to obtain and even harder to apply on any scale beyond a backyard operation, unless you have a tractor with a bucket. Note: horse manure can contain weed seeds.

Many urban growers can access quality city-generated compost these days for cheap. It should be noted that finished compost made into a liquid infusion (compost tea) and filtered very, very carefully through micron filters makes an excellent fertigation ingredient, as well as plant spray for foliar application. It supplies micronutrients, humic acid, beneficial bacteria and some NPK. It might be the best use of compost if your available quantity is very limited and you don't have enough to spread on the soil adequately.

To make compost tea, add 5 or 10 gallons of finished compost to a partially filled 55-gallon barrel of rain water, then fill with water, stir, and allow to sit 24–48 hours only, stirring occasionally. Filter through micron filters and use. An aerator is unnecessary.

Consider the impacts

Whatever fertility option you choose, consider the impacts of the source (who or what is being affected through sourcing the material) and try to source as local and regional as possible to avoid creating a high carbon footprint or unsustainable draw on resources. Some options, such as rock minerals, are non-renewable resources. But remember, you can always improve and change your fertilizer program later as more options are found and better sources become available. Be practical and utilize whatever appropriate organic fertilizers you can source, and refine and improve your sources later on, as you can. Because, in practical terms, you have to start somewhere. If you are not averse to animal products, chicken manure products are ideal for most growers and provide effective, complete, and rapid results, even when used alone.

Make sure any fertilizers are compatible with your organic status and local regulations and, if fertigating, make sure the irrigation system has a legally acceptable back flow preventer installed so fertilizer does not flow into the water supply you use. Follow all recommended dilution rates very closely.

Applying organic fertilizers

The first year in the ground, your fruiting plants' requirements for nitrogen are usually low and too much nitrogen can favor top growth to the detriment of good root system establishment. Too much nitrogen can also lower fruit production in general and favor vegetative growth, so go easy with nitrogen fertilizers.

Never just blindly add amendments you perceive as potentially beneficial, but do so only after careful consideration of your specific crop and soil conditions, usually in conjunction with a soil test or other regional recommendations. However, adding lots of compost, a little kelp powder, and a small amount of a balanced NPK organic fertilizer will go a long way to good crop health and establishment, in lieu of scientific precision. Personally, I never use soil tests, I just observe the crop and soil health, but it takes many years of growing crops and practice to be able to do so correctly. That may not be the best option for you, especially if you are just starting out. Initially, add organic insoluble fertilizers shortly before you plant the crop, by broadcasting it or applying evenly across the entire bed, and gently water in. When growing in rows, you can apply fertilizers to the soil around the plant shortly after transplanting. Never add insoluble fertilizers into the planting hole itself.

The second year, most fruit crops will benefit from the addition of nitrogen in the form of a modest spring dosage of high nitrogen fertilizers: poultry manure or corn/soy/legume-based products, etc. For a 25' (7.6 m) bed of strawberries I use about 10 lbs. (4.5 kg) of granulated chicken manure (4-7-4) and try to apply about the same amount of compost. For most larger fruit plants, around 2.5 lbs. (1.1 kg) of chicken manure per ft. (30 cm) of bed should give strong results. But you will have to explore your soil type and site and watch plant growth. In general, with good to excellent soil fertility plants should be lush, very green, have abundant foliage and flowers, and be setting a good crop of fruit. If the foliage is yellowing, sparse, or shoots are short and stunted, you urgently need more nitrogen. For large shrubby plants and blackberries about 1.5–2.5 lbs. (0.7–1.1 kg) *per plant* of chicken manure or high nitrogen fertilizer with P and K should suffice. I also aim for one or two foliar spray applications of diluted kelp extract every spring, to correct any possible micronutrient deficiencies and raise *brix*.

Do not fertilize any plants until your local frosts and freezes are complete! Even if they are budding out, wait! (This is mid-April in our region.) Otherwise, you may stimulate a strong flush of tender new growth that then gets destroyed by frost. This will at best set the plants back slightly and at worst it may lessen or eliminate the crop for the season.

In addition to soil testing, foliage nutrient testing is also gaining in popularity. You can send a few fresh leaves, bagged, to a testing lab for nutrient analysis. They'll send you a report describing if micronutrients are low, etc.

Perhaps even better than applying fertilizers, cover cropping with legumes before planting fruit plants can also provide cheap and very effective nitrogen availability to the crops the first season of establishment, as well as increasing organic matter and moisture retention of the soil. Check with your local ag extension office for local recommendations and timings.

Unlocking latent soil fertility

There is also a different school of thought, more along the lines of Masanobu Fukuoka, that holds that adding organic fertilizers is not necessary, and potentially harmful. We want to avoid fanaticism and absolutes, especially when they lack substantial scientific research and data to back their claims. Always remember that. However, the fertilizer-free camp does have some interesting theories and practices worth mentioning and discussing.

The theory of the fertilizer-free growers is that the soil already contains all that is needed to grow great crops. They say, "just look at the giant trees and healthy wild plants that receive no care, no fertilizer, and no tilling." They argue that what is needed is soil life and mulch, in order to "unlock" the latent nutrients and enhance plants' access to them. This is a great point and it's worth considering. There are many proponents out there and books on this topic you can look into.[2]

Basically, the fertilizer-free practices revolve around avoiding tillage and plowing and instead making prolific use of organic mulches like straw or wood chips. Often very important to this approach is the application of liquid inoculants that stimulate soil life. Liquid inoculants are primarily produced on the farm. One such soil inoculant that we have utilized comes out of India and is called *Jivamritam*[3] or "Soul Nectar." It is a rich, fermented liquid brew full of beneficial microorganisms, nutrients, and trace minerals. A recipe for making Jivamritam can be found in the Appendix, and there is a link to a video we made in the notes for this chapter.

Summary

Lots of excellent organic fertilizer options exist. Utilize what fits into your moral code, local/regional availability, and has the lowest carbon footprint. Fertigation with water soluble fertilizers should be highly considered. Don't overdo fertilizer applications or plant health will decrease. You may want to soil or foliage test before planning your fertility program. Consider utilizing the benefits of cover cropping.

6

Planting Successfully

Planting is something that seems simple but is often done very incorrectly and that is a costly error. Before we get to planting though, the first thing you need to know is how to properly handle your nursery stock when you receive it.

Handling and inspecting nursery stock

When purchasing from a mail-order nursery, coordinate the timing of shipment *before* purchasing, to make sure the plants will arrive when you need them, and track the delivery. Some nurseries may not be able to be super specific with shipping dates, so make sure to work with them as best you can to assure the plants will arrive in good time to plant, and when you expect them. For instance, are they shipping "in spring" or "in April", or on April 10th? Keep in touch and get tracking numbers. Don't be too worried about the exact shipment date, or hassle the nursery, just coordinate it to coincide with your regional planting period, which is usually *before* the last frost date. Freezing and frosty weather will not harm your dormant, leafless plants once planted in the ground, as long as they are not leafing out or budding.

When the shipment arrives, open the box immediately. Inspect the roots of the plants. They should be wrapped up tight in plastic and covered in a moisture-retaining material, often shredded paper, peat moss, or a moist gel substance. The roots should appear slightly moist but not wet, with no mold, rot, or foul smell. *They absolutely should have arrived that way.* Never let the roots dry out completely. If the roots appear slightly dry, spray with a fine mist of water, allow to drip for a moment, then remoisten the packing material and re-wrap the roots firmly, or soak in water for an hour, then repack as just directed. Do not soak raspberry roots.

If the plants arrived with very dry or otherwise damaged roots, clearly photograph it and contact the nursery right away and calmly describe the situation. If the roots appear totally dried out upon arrival the plants are likely to fail. Soak the roots in water or preferably kelp solution for 1–2 hours and re-wrap as described, just in case you're stuck with the shipment. Sometimes nurseries want the damaged shipment returned, which in my opinion is, in general, a poor business practice, except in special circumstances.

	Inspection Checklist
○	Are all the plants I ordered in the box(es)? Is the order correct?
○	Are the plants in good shape or are they broken or damaged?
○	Are the roots covered in moist material of some kind and wrapped in plastic?
○	Are the roots looking moist and healthy (or do they look dry, shriveled up, or moldy)?
○	Is the order clean of insects, slugs, and mold on the plants?
	If the answer is "no" to any of these questions, contact the nursery ASAP.

As you inspect the plants, a few snapped roots or twigs are no issue. See specific plant sections in this book for detailed guidance on handling and what to look for in quality stock.

If the answer is "no" to any of the inspection questions you should take several very clear photos and contact the nursery immediately. Email them the photos and politely and calmly let them know there is an issue. Don't freak out or be rude. These days shipping delays and problems can occur that are out of the realm of responsibility of the nursery. A reputable nursery will handle any situation responsibly, honestly, and expertly and you should trust what they say about the plants and not panic. A little mold on the roots is usually not a big deal, nor are some types of minor damage, such as broken twigs or a snapped root.

Did you read over the nursery's Terms and Conditions for warranty or for damaged shipment replacements before making a purchase? If you did not, and the plants are *not* warrantied or replaceable, then you will have to accept the results of your actions and do the best with what you've got. Always read a nursery's Terms and Conditions before making a purchase, especially a large expensive one. Research!

If the answer is "yes" to all the checklist questions, then unwrap the roots, gently mist with water or dunk in water and allow to drip for a minute or two and then re-wrap tightly in moist material and store correctly or, otherwise, plant immediately.

Storing dormant plants

If you absolutely cannot plant for a week or more, you will need to keep the plants completely out of sunlight and also freezing temperatures, ideally in a cold storage unit, root cellar, or large fridge.

Keep the plants as *cool and dark* as possible until ready to plant. Roots must at all times be wrapped up in moist material and plastic wrap or bagged. In cold weather, an unheated garage works very well for short term storage (1–2 weeks) as long as it does stay below 50°F (10°C) and does not go below 32°F (0°C). *Do not let the plants freeze.*

Heeling in dormant plants

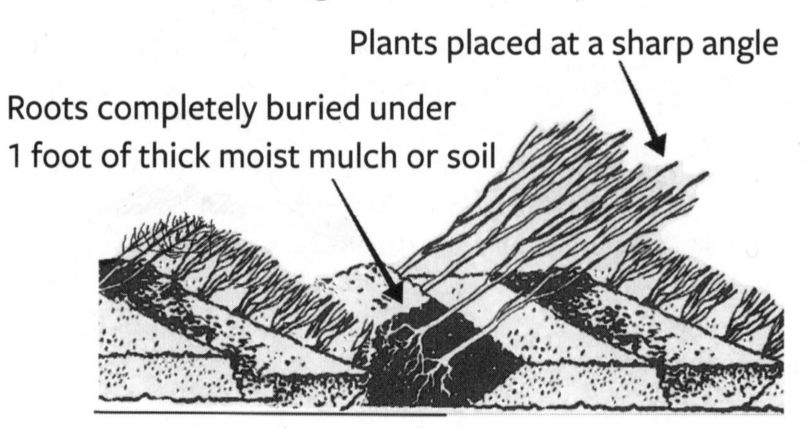

"Heeling in" works fine for short intervals of no more than a few weeks to a month in cold temperatures.

Heeling in is simple: You *unwrap* the roots and bury the roots under about 18" (46 cm) of heavy moist wood chip mulch, compost, or soil (*not manure or straw*). In cold northern areas bury under 2–3 ft of mulch/soil or store indoors. You don't want the roots to freeze, if possible. Placing the plants at a sharp angle (as if they are blowing over) will delay budding. If possible, do this in a shaded, cool area like the north side of a building. Keep the pile moist and when ready to transplant, dig the heeled in plants up very carefully, cover the roots in moist burlap/newspaper and plastic wrap, etc., and plant ASAP. If weather is above 50°F (10°C) they will only successfully store this way a couple of weeks at most. Tight green buds showing on the twigs means your time is up and the plants have to be dealt with immediately or they will suffer transplant shock and may have reduced shoot growth the first season (but should still survive at least and recover by next season).

At ambient room temperature (65–75°F or 18–24°C) dormant bare-root plants will only stay viable about a week or so before they begin to bud and experience stress. Chilled to 34–39°F (1–4°C) and kept well packaged with moist roots they can be stored for 2–4 weeks in excellent condition.

This all goes to show how important proper timing is and having everything lined up well before you start. Ideally you would have plants arrive at a good planting time, store for no more than a few days, and plant. Sometimes rain or even snow can delay planting by a week or more, so be prepared.

What if the roots freeze? For example: the shipping box accidentally sits outside all night in freezing weather. If the roots or plants briefly freeze, leave them wrapped up and bring them in to an above freezing, but not hot environment. Allow to sit in the warmer temps (40–60°F, 4–16°C) for 24 hours, then keep cool and plant ASAP. Likely they will be fine as long as the roots stayed moist and the plants were not in freezing temperatures very long (12–24 hours).

Effectively handling cuttings

Cuttings should arrive bagged tightly in plastic and may or may not have moist material added, but should be sealed up and appear very fresh and moist. Greenwood cuttings are those taken during active growth of the plants when the stems are green and flexible and not yet lignified into wood. Keep wrapped in plastic, add a small drop of water or a very brief fine mist of water inside the packaging for humidity, and keep refrigerated (but non-freezing). Use within 1–5 days. Plant them immediately in well-draining potting soil, under intermittent mist or in a humid chamber, both of which are beyond the scope of this book to describe.

Dormant cuttings and root cuttings should also be free of wrinkles and appear fresh and vital with tight buds. Removing a tiny section of bark off the cutting should reveal a bright green inner cambium layer, which should not appear brown or yellow. Any cuttings should be free of mold and rot and should not smell foul or be soft. Dormant cuttings and root cuttings should be kept humid and wrapped in plastic. An exception is fig cuttings which mold easily and so should be kept dry, with no added humidity, sealed in plastic or in a ziplock bag.

Make sure everything is carefully labeled and does not get mixed up. Keep all cuttings refrigerated at 34–39°F (1–4°C) (but non-freezing). Use all types of cuttings as soon as possible to avoid them spoiling, drying up, or rotting in storage. Be prepared to use the cuttings when they arrive.

Pre-planting preparations

Being ready to plant means the site is prepared, your tools and supplies are on hand, work crew (if utilizing labor) is ready, and the weather and timing is right. Good planting weather is calm, not hot or intensely sunny, and not windy. Overcast and calm is perfect. The full moon or several days before or after is ideal but not necessary.

Excellent pre-planting prep includes inspecting the plants, doing any quick pruning of broken roots or twigs, soaking for 30 minutes to an hour or two in kelp solution (or water, either in buckets or tubs). Don't soak roots more than 1 or 2 hours or damage can result. Don't soak raspberry roots as this can damage them. You can utilize pre-mixed kelp extract solution or, even better, use one tablespoon of solid kelp extract to 5 gallons of water and mix thoroughly.

When planting, *it's important to make sure roots stay very moist and covered 100% of the time.* Soaked, used burlap bags or burlap cloth works perfectly to cover plant roots while planting. Keep roots covered and never openly exposed to the desiccating sun and wind. Wet sheets of thick newspaper work too, or keeping roots in buckets of water, but for no more than a couple of hours. Any plants left unplanted should have the roots sprayed with water, then bundled in moist material, wrapped in plastic, and stored out of the sun somewhere cool (or put back into cold storage, if available.) Plant leftover plants promptly, or if they're extras, try quickly selling them off or potting them up. Don't leave them exposed to die or in buckets of water overnight; it may kill the plant. Make sure plants stay labeled/tagged or organized throughout the planting process—you don't want to find later that you mixed up the cultivars, as this could become a serious problem.

Proper planting practices

Over the years I've witnessed and observed all kinds of ineffective and incorrect ways to go about planting fruit trees and berries. Let's set the record straight.

First, *never put anything in the hole at planting:* no peat moss, no fertilizers, no manure, no buckets of compost, no Miracle-gro, no potting soil, etc. Nothing needs to go in the hole except the soil that was just removed, the plant's roots, and water. A little compost and crushed rock minerals (rock phosphate, greensand) is OK but not necessary. Blueberries usually require the addition of peat moss to the planting hole. See the section on Blueberries in Part 2 for details. Adding anything else can cause harm and damages the plant.

Small fruit plants are preferably planted by hand and not with some kind of machine unless you are planting a great quantity or are physically unable to do so. Augers, mini-excavators, etc., are usable but can disturb the soil and dig up weed seeds and tear into beds, so it's better to do it by hand using a sharp flat-bladed spade (not a rounded scoop shovel, which is designed for moving loads, not digging holes). Make sure the spade is sturdy and sharp. For small plants or in very soft soil, a Japanese hori hori (garden knife) can be very useful.

Planting holes should be big enough to accommodate the entire root system easily without cramping or wrapping roots—never cramp the roots! I find 18 inches wide (46 cm) by the depth of the root system of the plant is a good general size for larger plants, trees, and vines. Small plants (strawberries, tomatoes, raspberries) can be rapidly planted with a trowel or hori hori garden knife, or use a hoe to rapidly pre-dig evenly spaced holes and then just pop the plants in. You could use a seedbed roller with markers attached to quickly get the spacing exact.[1]

When digging, separate the top soil from the subsoil and when refilling the hole, put the richer top soil back into the hole first.[2] Quickly strike and break up the sides and bottom of the hole with the spade. Take the plant in hand and place it in the hole so the roots begin just below

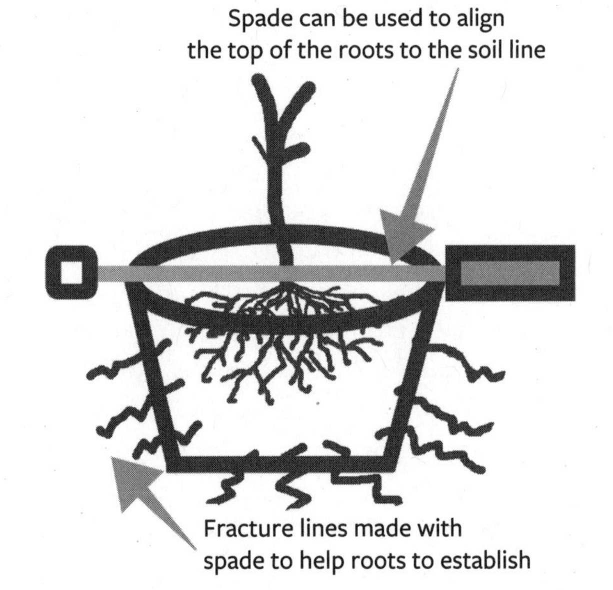

the top of the hole—in other words, do not bury the stem of the plant. For larger shrubs and vines you can place the spade over the hole and use that to help gauge what level the stem should be at.

As you plant, continually spread the roots out like a wide open hand; they should be going in all directions and not cramped together. If you have very long roots that are impractical to dig holes for, cut them back to 18 inches. Sometimes blackberry and raspberry roots can be like that—cutting them does not harm the plant. If desired you can dig trenches for long rigid roots or roots near the surface and place them within the trench. A trench can be rapidly created with a spade by striking it in the desired spot and prying it from side to side.

Carefully refill the hole and gently pack down the soil as you go by pressing the soil with your hand. Make sure the stem of the plant is not sinking below the soil line. Better to plant slightly higher in the hole than too low. Once the hole is filled, tamp the soil down gently with your foot—*no stomping!* Then immediately water deeply with about 1 gallon (~4 l) of water or, preferably, kelp solution (which helps with transplant

Kelp Solution

Add 2–3 tablespoons of solid kelp extract (I use Kelp-It brand) to 5 gallons of water. Allow to sit for 15 minutes and stir thoroughly before using.

Or, add liquid extract to water as per package instructions.

Soak plant roots in the solution for 1–2 hours or use to water the plants.

Use solution within 2–3 days.

stress). Small berry plants require less water at planting: around a quart (1 l) of water each. Strawberry beds can be soaked entirely at once after planting.

Sometimes soil in the hole will sink or compress down when watered, especially if not filled or packed properly during planting, taking the plant downward deeper into the hole, and you don't want that to happen.

Post-planting care

Absolutely ensure everything gets immediately and deeply watered, no matter how tired you are! You can also plant right before a heavy rain storm. A light rain of less than 1 inch will not adequately water in plants and can lead to failure.

If you notice the plants are sinking down into the holes, adjust your planting depth and make sure you're backfilling the hole and tamping it in firmly enough. Add any support stakes, labels, etc. at this time. Make sure the cultivars stay labeled or organized as needed.

If you are planting bare-root canes of blackberries, blackberry hybrids, or red/black raspberries, now is the time to *prune the canes to the ground*, leaving no stump, only the roots. Clip them with snips. This prevents the plants from producing a few unnecessary fruits the first season and diverts the plant's energy into establishing strong roots and getting off to a more vigorous start. When planting small plugs, *do not prune at planting.*

If your plants are *small plugs or are in small pots*, simply pull back the soil with a hori hori or trowel and place them in, then re-cover and water in.

After planting and watering, cover with any mulch or landscape fabric as soon as possible. This will help deter weed growth and keeps the soil moist and protected.

Summary

Make sure to handle your bare-root plants or cuttings properly and carefully. Planting correctly is crucial to the future success of the plants. Take your time and do it right. Hire help if you are planting a lot. Make sure everything is organized and lined up, including the site, labor, tools, etc., before plants arrive. Plant in good weather conditions. Follow specific planting instructions per plant species.

After planting, perform post-planting care: laying mulch, installing stakes or trellises for canes/vines/tomatoes, applying compost/fertilizers, and any cages for plant protection—see Chapter 9: Maintenance, Protection, and Pest Control for more details.

7

Organic Weed Management

Managing weeds effectively and organically is an important factor for success. If you've ever heard about organic farmers pulling weeds all day long, this is because of poor weed control strategies (or trying to do too much). You can't afford to be pulling weeds all day, it's inefficient and it's also hard on your back and your morale. Therefore, effective strategies of *prevention* and also *quick removal* are necessary. Remember: berries and fruit trees, especially when small and establishing, cannot compete with grass or weeds of any kind within the planting bed. During establishment, fruit plants can rapidly get overtaken and permanently stunted or killed off by the competition. Don't plant more beds or rows than you can maintain effectively. Given time, this will correct itself (if you overdo it initially) as you lose control over some of the area.

In this chapter we review some effective strategies.

Using permanent grass cover

For maintaining the area between beds or rows, most orchards and small fruit farms utilize grass in between, and mow it regularly. This has pros and cons. Grass protects the soil from erosion and is better than rows of weeds. A non-spreading grass such as orchard grass, bluegrass, and fescues work well in many areas. *Avoid creeping or rhizomatous grasses* such as quack grass, Johnson grass, or zoysia, which will rapidly creep into the planting beds and lead to major never-ending weed problems.

For mowing on our farm, we use a European BCS walk-behind tractor with a 3' (1 m) mower attachment, which allows close mowing and close row spacing, due to the walk-behind's small size. Honestly, a zero-turn or riding mower would be much easier. Utilizing a small mower can make the layout of an orchard/market farm more space efficient because beds or rows can be created closer together, with less space dedicated to wide, tractor-spaced grassy paths. Keep it cut: tall grass encourages the proliferation of crop-destroying insects, mollusks, rodents, and weeds.

If you want to avoid grass, excellent *permanent ground covers* include ladino clover (also excellent honeybee forage). Investigate effective local and regional options before deciding on ground covers. Many non-rhizomatous grasses and ladino clover can grow together. Weeds *will* grow in ladino clover, so it will require some mowing and reseeding every few years. Grass and clovers also are effective mixed.

Some people might appreciate experimenting with *animals grazing in between rows (silvopasture)*. We have successfully used moveable electric netting to graze dairy goats in between tree fruit rows. It can be a lot of work moving the nets, but some people enjoy it. Permanent electric fencing could allow cows or other grazers (even small ducks or geese) to do their thing safely in between rows without destroying fruit plants. In a system such as this you could be simultaneously farming multiple products besides fruit (milk, duck eggs, wool, etc.). There are many ways to manage the grass growth besides using fossil fuel resources, and your time, to suppress grass—an endless perennial drain on both fronts. What *products or useful purposes* could these areas potentially yield? Perhaps organic hay, honeybee forage, pasture? Whatever your choice, you must design carefully and dedicate yourself to maintaining the rows. (While not ideal, mowing is still a relatively easy and simple way to manage grass and pathways, keeping it all in check.) Fully describing alternative systems is beyond the scope of this book, but such descriptions can be found in permaculture and sustainable farming publications on silvopasture. It may only be realistic to graze ruminant animals in between larger fruit plants such as grapes, mulberries, etc. as the pathways would have to be sufficiently large to accommodate them.

Mulch[1]

Organic mulch can help optimize soil health. Ideally, all trees, berries, and shrubs would have a constant layer of decaying organic matter such as thick straw, deciduous wood chips, etc. Imagine the rich humus layer in a healthy forest. Organic mulch should never touch the trunks or stems of the fruiting plants, as it can lead to rotting of the stem. It should also not be more than 5–6 inches (13–15 cm) thick or it can deprive the roots of oxygen.

Overwintered bed showing 4 different options. Covered bed is strawberries being protected from late frost.

Many organic growers are going to want to use organic mulch. Organic mulches give the best biological results overall and enhance soil health. Straw, wood chips, grass clippings, dried leaves and leaf mold (decomposing leaves), all work well. If using fresh bark, sawdust, or wood chips, add extra organic nitrogen material on top of the soil (under the mulch) or liquid feed nitrogen in the form of water-soluble soy/corn fertilizer. This feeds soil bacteria, thus preventing them utilizing nitrogen in the soil to help breakdown the fresh woody material, which can lead to nutrient deficiency in the crop. Straw is often expensive ($5–8 per bale these days), and if not organic may contain traces of glyphosate (Roundup). If you can get it in bulk for cheap, go for it. It makes a near ideal mulch material.

The main *advantages* to organic mulches are that they are natural environmental or agricultural byproducts and so, as they break down, they supply abundant organic matter and add some amount of nutrients and trace minerals. Mulch stimulates soil micro-life, beneficial fungi, and feeds earthworms. Big advantages indeed! Also, in various forms it is often available free of charge, and has no toxic residues (unless treated or contaminated with chemicals). It helps conserve soil moisture and protects the soil from wind and water erosion, and keeps it cooler on hot summer days.

The main *disadvantages* to organic mulch materials are (1) they are sometimes difficult to obtain in quantity, (2) they are often extremely heavy and labor intensive to apply and transport, (3) they often rapidly break down and need annual reapplication, (4) weeds readily establish within the rotting mulch layer and can be difficult to remove, and (5) they can sometimes contain litter and traces of pesticides such as glyphosate or fungicide sprays. So, be careful where you source from.

An excellent and usually contaminant-free source of wood chips is any local or municipal tree-trimming operation. Be forward and politely inquire with the local companies if they would be willing to dump chips on your land. When I say be forward, you may have to flag down trucks or visit them on the spot when you see them cutting limbs or trees down. Be brief and to the point and have your address written down clearly on paper. Make sure you have a dumping site that is convenient, safe, very near your plot, and accessible by an enormous, top heavy, and extremely heavy truck. If the site is overly wet, sloped, or risky for a huge truck, don't expect to get chips delivered there. Also, be sure to be very communicative about when and where and how many loads you want dumped. Once they find a good and free site to dump chips, they can become overzealous! Beware of chips containing large thorns that can pop tires, such as from honey locust trees and Osage orange. Also, strictly avoid any chips containing black walnut, which contains chemicals (*juglone*) highly toxic to many other plants.

To remedy one of the main disadvantages of organic mulch materials, that is, their rapid decomposition in wet climates, a thick layer of flattened, plain brown cardboard placed under the mulch layer will extend the usefulness and life of the mulch considerably, and kills most of the grass and weeds underneath. Make sure the cardboard is covered in organic matter or it will blow away. Also check with your Organic Certification rules if you are or are considering getting certified, to make sure cardboard or newspaper is allowed in your production system. Cardboard can be obtained for free in large amounts from industrial centers, grocery stores, factories, and many other stores. Inquire.

Semi-synthetic and reclaimed mulches

This is a bit of a "grey zone" and may not be approved for organic certification, but is highly worth mentioning because we live in a world of grossly excessive resource use and there are millions of tons of random materials lying around every human-occupied place. Some of these random materials make potentially effective weed-smothering materials for pathways: old carpets, used plywood, old sheet metal (covered in gravel or mulch), flattened plain cardboard, used greenhouse plastic—the list goes on and on. I would investigate where it came from and only utilize non-contaminated and pesticide-free materials. Also, many of these could eventually shelter slugs, snails, ants, and pill bugs underneath, so be careful.

In Germany I explored a small chemical-free greenhouse operation utilizing old plywood sheets in the walkways, and wood mulch everywhere else. The greenhouse was highly productive, weed-free, and pristine. These materials are also FREE.

Current *bio-based plastics* are also semi-synthetic, and many only contain 10–20% bio plastic from corn, the rest being synthetic polymers. These are banned from Certified Organic agriculture in Canada. Make sure whatever you utilize will not degrade into micro plastic and contaminate your soil.

Landscape fabric

The other option is *heavy duty, woven landscape fabric*. Avoid spun material or those that tear easily. Always keep the soil around fruit plantings covered, even if only using synthetic material such as woven plastic landscape fabric.

These are the main *advantages* to landscape fabric: it blocks weed growth extremely effectively; installation is fast and easy, especially when compared to moving ton after ton of organic mulch; its durability is at least a decade. Compared to organic mulch, the labor reduction with landscape fabric is tremendous, not only because of its easy initial installation but because there is no necessity for yearly re-application. For those with any form of disability or labor shortage, using landscape fabric will make the most sense.

The main *disadvantages*: the synthetic nature of the material and its environmental impact, including the need to eventually dispose of used materials. (One possible future option to reduce this impact is using bio-plastics, which continue to rapidly evolve and become better and cheaper, hopefully one day being 100% bio based.) Also, synthetic mulch does not add any nutrients or organic matter to the soil. These disadvantages are deep but will have to be weighed realistically against your goals to feed your community and make an honest living while making your operation sustainable for you, your labor force, and your sanity. Choose wisely.

Another option is utilizing synthetic mulches when necessary and organic mulches or permanent ground covers as you are able. It doesn't have to be totally one or the other. For instance, you could use synthetic covers in walkways, and organic mulch on your growing beds. This is a great option.

Landscape fabric can be cut to size and placed around plants, or stretched down entire planting beds or walkways of any length or width. It can be held down securely by hammering 6-inch-long (15 cm) metal sod staples through the cloth into the ground. These staples should be used about every 18–24" (45–60 cm) of material.

Some make the argument that landscape fabric damages the soil and ruins it. That has not been my experience. It allows air and water to penetrate. It simply blocks sunlight and thus suppresses vegetative growth for a number of years, allowing the desired fruit plants to get a strong, unhindered establishment without resorting to herbicides. Underneath it you can apply compost, or fertigate with filtered compost tea, *jivamritam*, etc., to help stimulate soil life around your fruiting plants.

If you're determined to *only use organic mulches*, then that is your personal choice and I sincerely hope it works well in your operation. But in time you may find that sourcing it, keeping up with annual reapplication year after year and fighting the weeds growing in it, on a small farm scale, is unsustainable for you or is harming your yields. Understand that you have good options that do not include using toxic chemicals and herbicides, or require long, hot days on your knees pulling weeds and fighting a basically losing battle. On our farm both landscape fabric and organic mulch give good results and no observable differences with plant health.

When using landscape fabric, you can use either white or black. Black *warms* the soil and blends in visually better than white. However, white remains *cooler* and also reflects light onto the plants, which can increase their health and production. In extremely hot climates the white ground cover may perform better.

Installing landscape fabric

When installing plastic, *either cut individual squares* for each shrub/tree/vine, including large berry plants like blackberries and blueberries, or, alternatively, *just lay down a long line of it.* For blackberries, vines, trees, and shrubs, we prefer individual squares but, for most berries planted closely together, a long line gives the easiest and best weed and grass control. See individual fruit plant listings for details.

Install landscape fabric (and irrigation drip tape or lines) *before installing the plants*. Tractor-mounted attachments exist that lay the drip tape and plastic at the same time. Also, these can be obtained for walk-behind tractors.[2] Usually this is for temporary, one-season plastic cover used for annual vegetable production. Can the tractor mount be used for landscape fabric? This is worth looking into if you're doing this on scale. If doing it by hand, simply lay the fabric on the bed or site (on a windless day if possible!) and hold it down with some blocks or sand bags and then staple it down along the edges every 18–24". Burn or cut holes in the fabric to install your plants into. A propane weed torch works great. A soup can attached to the end of the torch, when heated, makes a clean circle in the plastic. This works very well for strawberries.

Make sure to cut out a circle around each plant with a sharp knife or scissors, so the cloth is not smothering it. The cut circle should be about 5–8" (15–20 cm) diameter—large enough to water or apply organic fertilizer, but small enough so that the weeds are in the dark and don't proliferate within the opening. This works great for blueberries, blackberries, gooseberries, and larger fruiting shrubs and trees.

Make sure when cutting holes for suckering plants such as blueberries, raspberries, or blackberries that they are large enough to facilitate *replenishing shoots and suckers* at the base or underground near the stem, and not get blocked by having the landscape fabric too close to the stem/trunk. If those shoots are blocked, the planting will fizzle out after a few years.

It's difficult to get the landscape fabric lined up perfectly with plants in-ground, so cut each opening one at a time as the fabric lays in the row, next to the row of plants. After each opening is cut, place the landscape fabric over the plant, with the plant inside the opening, and secure with staples on the edges of the fabric. For larger plants you will need to cut a slit so you can slip the landscape fabric around the base of the plant. Use wide plastic that is 3–4' (0.9–1.2 m) wide, often available in 100–500' (30–150 m) rolls. Drip irrigation lines are generally placed *under* the plastic mulch before laying it. When adding amendments after planting, put them in the hole around the plant, not on top of the plastic.

Be aware that if you cover the plastic material with organic mulch, either for aesthetic reasons or to try to stimulate the soil life, these materials will decompose *on top* of the plastic and weeds will in turn germinate *on top* of the plastic and will root *through it*, thus eliminating the benefit of it, and causing a mess that cannot be easily dealt with. So, it's better to keep it fairly clean.

Overall, using single squares of landscape fabric around each larger fruiting shrub/tree works great, looks better, uses substantially less plastic, and is probably the better option for larger operations.

Another very effective way to install ground cover is the *open bed method*. For this, the bed is left uncovered and completely open, and only the walkways are covered in landscape fabric. This works great for suckering brambles like raspberries, strawberries, and small plants, including tomatoes. Strawberries, raspberries and other heavily suckering plants need to be able to *easily sucker* from below ground or via tip-layering in order to keep the patch going, and if they were closely mulched with plastic those shoots would hit the plastic layer and dry up. Thus, keeping an open bed, yet covered walkways, is very strategic.

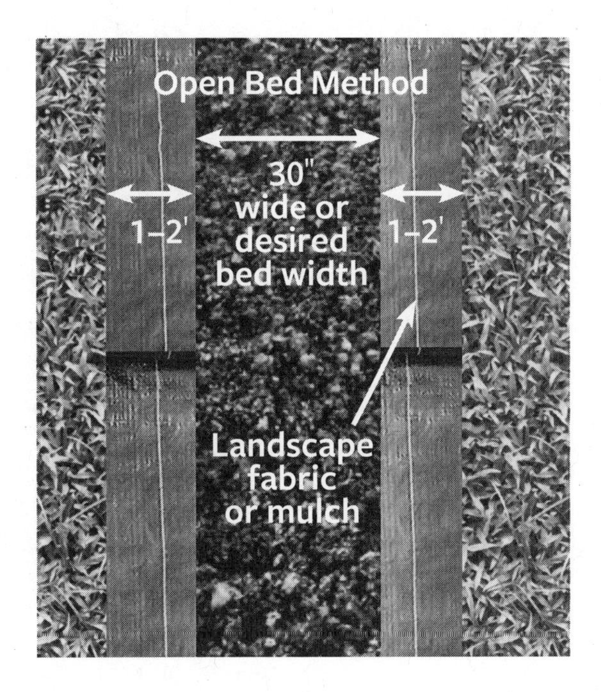

Flame weeding

Flame weeding is a newer technique that is rapidly gaining in popularity, even among mega industrial organic growers (who now have tractor-mounted flame throwers!). It's simple: first you completely prepare your growing beds. Then, water the bed if possible and allow to sit undisturbed for about 7–10 days. Thousands of tiny weed seedlings will likely emerge. While they are still tiny and fragile, sweep across the entire bed with a propane-fueled weed torch. Back-pack mounted units and also cart-pulled (as well as tractor mounted) options exist. Pass over the bed rapidly with the flame. A split-second exposure to the intense heat is all that is needed. You may go over it twice if it is a wide bed, as it can be hard to reach both sides. All the weeds will be rapidly killed off and you now have a "clean" bed. Immediately after flame treatment, plant your crop. If you plant carefully and limit other soil disturbance, very few weeds will emerge, and these can be weeded easily. This is an effective, modern, highly efficient, no-till method for organic weed control.

Lemon balm, sorrel, and ground ivy suppress weeds and do not interfere with gooseberry (in background)

Plant Guilds

Finally, you can also implement a *guild* sort of arrangement where low-growing, weed-suppressing perennial plants are established around your larger fruit plants. Examples of effective guild plants are: lemon balm, mints, chives, garlic, ground ivy (*Glechoma hederacea*), sorrel, marigolds, clovers, etc. (I wouldn't recommend comfrey as it gets too big and will interfere with most small fruit plants, but it is ideal around trees.) These may have to be occasionally cut back, yet can provide useful harvests of their own, and many will attract pollinators. Also, many small fruits such as gooseberries, rapsberries and currants make good guild members for larger fruit trees.

Determinate tomatoes grow in plastic pots in this efficient Amish setup. Note the support rods attached to the ceiling purlins and the very close spacing.

Aronia are a hard sell but have potential as a processed fruit. 'Mckenzie' aronia produces copious yields of no-spray, super-high anti-oxidants and outstanding flavor once processed (and sweetened).

Credit: Honeyberry USA

Small metal hoops (likely EMT), draped in bird netting allow lower-growing fruits like currants, strawberries, etc. to be protected.

Credit: Blue Fruit Farm

Blue Fruit Farm in Minnesota effectively markets honeyberry, aronia, black currants, and other 'unusual' small fruits. Here bird netting on tall supports keeps large fruit bushes safe from birds.

Blackberries are easy to grow, very productive, and the new University of Arkansas releases have upped the quality, sweetness, and consistency of this crop.

Superior cultivars of mulberry, such as this Thai strain, show astonishing productivity, large berry size, delicious flavor, and real marketability. A 'fruit of the future' in the West.

Red raspberries, like these 'Caroline' berries, remain one of the best potential berry crops for the market grower. They will not thrive in regions with extremely hot, humid summers.

An excellent fig for cold hardy high tunnel production and marketing is 'Chicago Hardy'. Here pictured at peak ripeness (with visible cracks), yet too soft and perishable at this stage for marketing purposes.

Instead of suppressing growth under small fruits (as with mulch/landscape fabric/etc.) you can often create a 'guild' or team of low growing, non-vining beneficial and/or marketable plants. Here, lemon balm, sorrel, and ground ivy grow under gooseberries.

Maintaining strong popularity in Europe, currants, like these reds and blacks, are a very solid market crop where popular. Note the wooden berry boxes used in this photo taken in 2021 in Germany.

Adorable as they are resilient, the humble bumblebee (here on gooseberry) will pollinate most small fruit blossoms, even in cool, wet weather when honeybees are loathe to leave the hive. Protect them.

Passionfruit (*Passiflora incarnata*), a native USA fruit, holds strong potential for becoming a new fruit crop. This superior strain is maintained on our farm and is prolific, with large, flavorful fruit.

Black raspberries like 'Hidden Gem' are delicious fruit but more challenging to produce in quantity and to market than reds.

Cold hardy figs grown in unheated high tunnels hold promise as a productive niche crop for market growers in areas too cold to effectively grow figs outside. Author's high tunnel in central Kentucky pictured.

Honeyberries, or haskap, a northern berry crop well suited to organic production, are rapidly growing in popularity. They could be a good niche crop in Canada but as yet are not a popular fruit in the USA.

Japanese beetle, a voracious, invasive insect, feeds on the foliage of many fruit crops (here, gooseberry), and consume blackberry fruit. Challenging to manage, but pyrethrins work well.

Morus alba mulberry fruits come in an array of tones of black, purple, lavender, pink, and white, furthering their potential appeal. Light colored fruits like this 'Honeydrops' mulberry are milder and non-staining!

This freshly hatched lacewing larva wastes no time, and immediately starts to hunt aphids on this grape leaf. Lacewing eggs are a great investment in natural pest control.

This super productive and delicious Juneberry was discovered growing in a parking lot in Louisville, KY. Netting is usually required for bird protection, but here birds apparently hadn't found them.

Organic semi-determinate hybrid tomatoes growing in-ground in the author's unheated high tunnel. Note drip tape, open-bed method, and head lettuce (mostly harvested) growing underneath. Twine keeps them upright, tied to purlins. Tomatoes sold for $5.00/LB.

Epic production, great quality fruit, and easy organic cultivation make muscadine grapes a winner in the South, and warmer mid-Atlantic areas. This 'Carlos' grape grown in the author's high tunnel in KY produced around 50 lbs of fruit.

Modern hybrids such as this 'Marnero' tomato combine excellent flavor and production with a colorful appearance. They are a much better choice for most market growers than heirlooms, especially in tough growing conditions, and often have as good or better quality and flavor.

Munson grapes, a fruit of the future developed in the 1800's. This 'Gold Coin' cultivar, a delicious table grape, exhibits extreme disease resistance, producing perfect seeded grapes with no sprays on our farm. Hundreds of cultivars perilously still exist in this line, waiting to be utilized before they disappear forever.

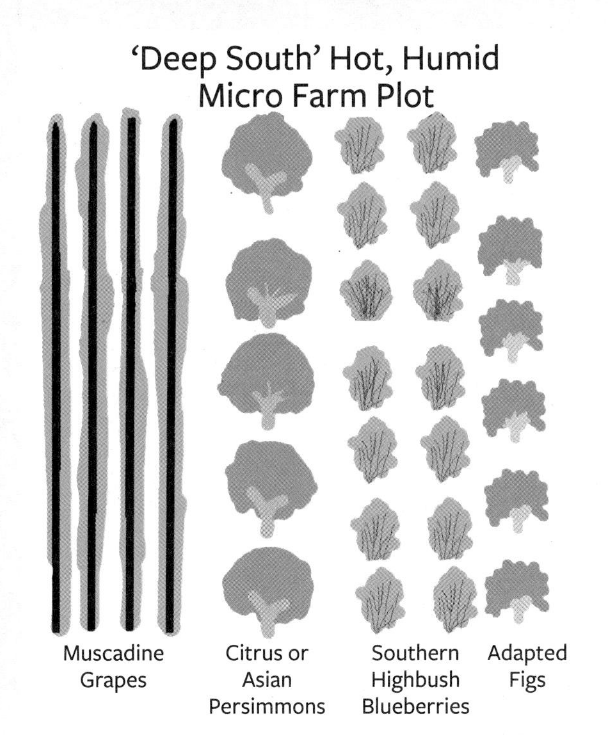

A sample plot for small marketable fruits and smaller tree fruits suited to a hot, humid climate like that found in the Deep South USA.

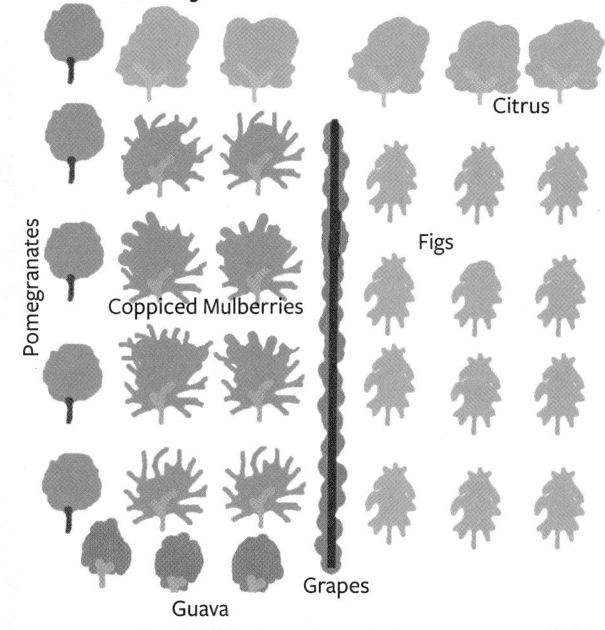

A sample plot of various especially easy to grow and profitable small fruit and tree crops in a warm climate.

Colorful bunches of cherry tomatoes in an organic greenhouse in Germany. Note the cool metal spiral supports, drip tape, open bed method with wood chip mulch, and old plywood used in the walkways.

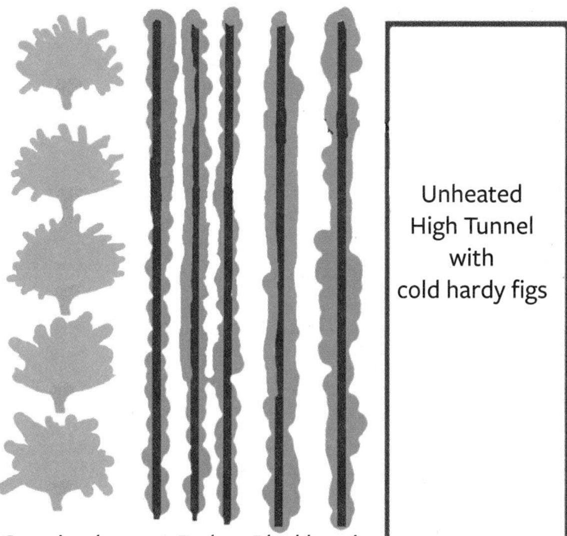

A sample plot of small marketable fruits suited to the Mid-Atlantic.

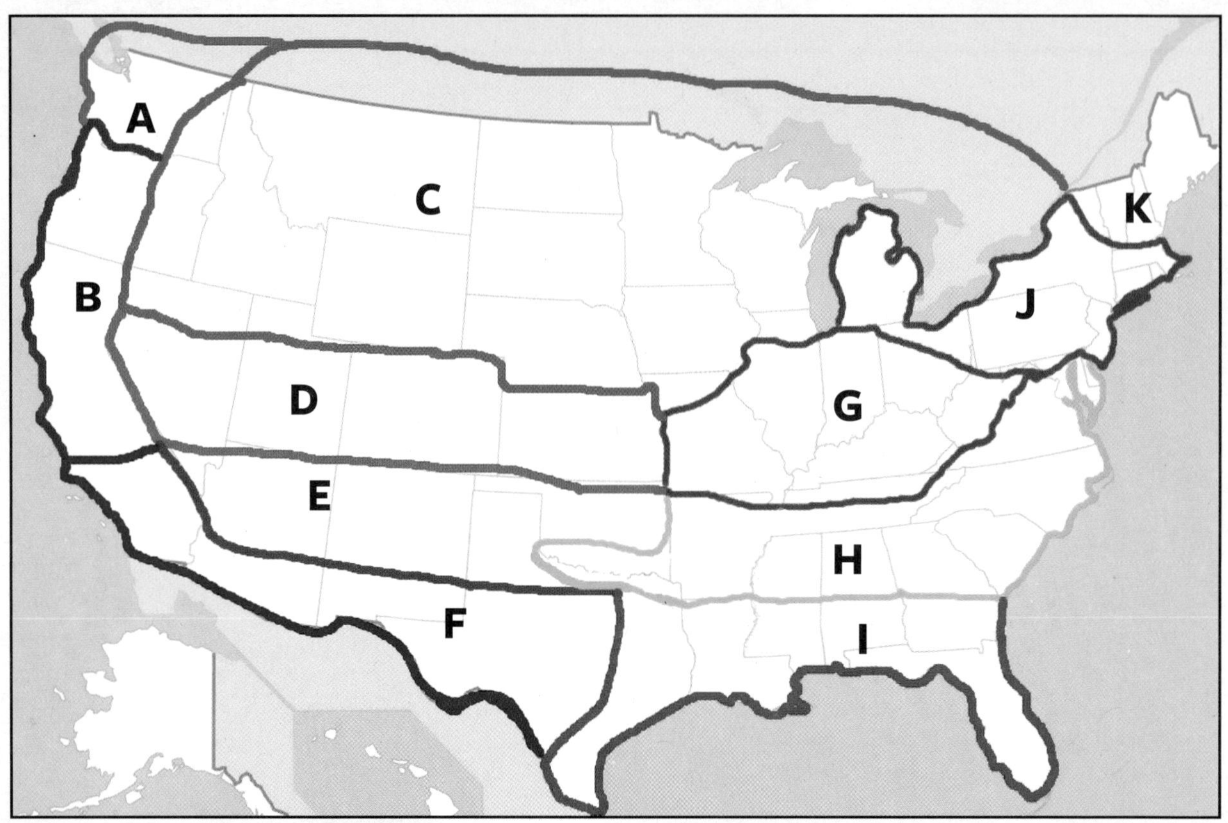

Small Fruit Growing Regions. **Note**: This is a very generalized map to what small fruits may be most successful for marketing purposed in your area. This is a starting point. Check with your local agricultural extension office and other regional growers before proceeding and make sure cultivars are adapted to your USDA hardiness zone.

A
Tomatoes, Berries, Adapted Figs, Blackberry

B
Tomatoes, Kiwi Berry, most berries, Figs, Pomegranate, Feijoa

C
Aronia, Hardy Berries, Saskatoons, June Berry, Honeyberry, Figs in tunnels in warmer areas, Cold Hardy Mulberry in warmer areas

D
Figs in tunnels, Hardy berries, Cold Hardy Mulberry

E
Hardy Pomegranate, Cactus Fruit, Hardy Figs, Mulberry

F
Jujube, Pomegranate, Figs, Mulberry, Blackberry, Cactus Fruit

G
Tomatoes, most berries, Figs in tunnels, Mulberry, Aronia, Honeyberry

H
Cold Hardy Muscadines, Tomatoes, Figs, Juneberry, Hardy Pomegranate, most berries, disease-resistant Mulberries

I
Adapted Figs, Muscadines, S. Highbush Blueberry, Feijoa, adapted Pomegranates, Blackberry, adapted tomatoes

J
Juneberry, Tomatoes, most berries, Kiwi Berry, hardy Mulberry, Honeyberry, adapted early-ripening Figs in urban microclimates or in tunnels

K
Hardy berries, Saskatoons, hardy *Roseaceae*, adapted tomatoes, Honeyberry

8

Tools of the Trade

Every traditional craft has its share of cool, extremely useful, very specific tools and small fruit farming is no different. Some of these are modern inventions, others are modern takes on extremely ancient tools.

Buy only the highest quality tools you can acquire. Cheap tools like the ones found at big box stores (or even many smaller hardware stores these days) will prove to be an exercise in futility and inefficiency and will frustrate and exhaust you. See the Resources section for sources of quality tools.

Hand Snips: In Europe these are called "secateurs." Also called hand pruners. Snips vary *a lot* by brand, size, and type. If you have small hands, make sure to buy smaller snips.

Buy a belt holster to securely carry them—and look very professional at the same time! Otherwise, they can tear a hole right through your pocket. This tool is dangerous and should be kept well out of reach of children. Keep them very sharp. Special diamond-edged small sharpeners work well. Don't use a typical file! Also oil them (food grade oil only) and keep them stored out of the weather and clean. A steel-bristled brush works well to clean off gunk, rust, and plant sap.

Snips are used to prune berry bushes, figs, tomatoes, and trees, and also to harvest crops like grapes. Special blunt-edged snips are available for grape harvesting. You'll be using snips all the time, so get an excellent quality pair, a holster, oil, steel brush, diamond sharpener, etc.

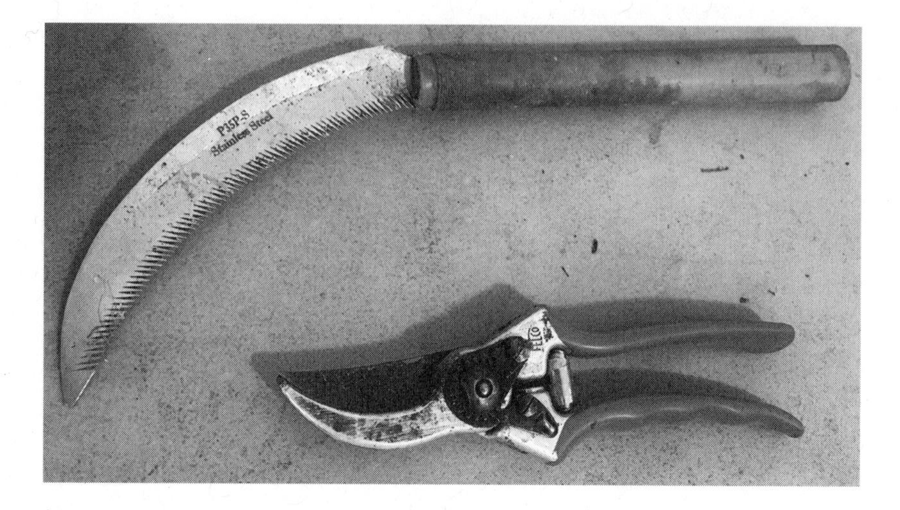

Kama weeding tool on top. Highest quality Felco brand snips on the bottom. Both get very regular use on our farm.

Loppers: Basically, larger hand snips with long handles, loppers are for cutting larger stems/branches, over about ½ inch (1.25 cm) in diameter. These are for removing spent blackberry canes, pruning thick stems, taking out weed trees, etc. Used enough that you'll want good ones. Same care as with hand snips.

Kama: Once you use this little Japanese tool, you'll wonder how you ever gardened or homesteaded without it. It's basically a thin, crescent-shaped, serrated steel blade on a straight 8–10" (20–25 cm) handle, usually made of plastic or wood. You rip through weeds and vines with it. For instance, you grab a handful of weeds or vines with the left hand and pull the kama with the right hand, easily slicing through them in a split second. An amazing tool for rapidly taking down weeds—very efficient and easy to use. Again, somewhat dangerous, so be careful with it and keep it away from children.

Hori Hori: Also called a garden knife. This increasingly popular bowie-knife looking Asian tool is extremely versatile and useful. It can be used for digging out tap-rooted weeds, slicing twine and weed vines, and rapidly digging small holes to plant things like strawberries and tomatoes. An essential planting tool. Holsters are available. Again, use and store with care.

Tapener: This handy device is like a large staple gun that wraps a short length of flexible tape around a vine or plant, and when squeezed, staples the tape together, securing it to whatever it is next to it (stake, trellis, etc.). This is very useful for staking and trellising berries and grapes. It is much more efficient and faster than trying to hand-tie plants with twine.

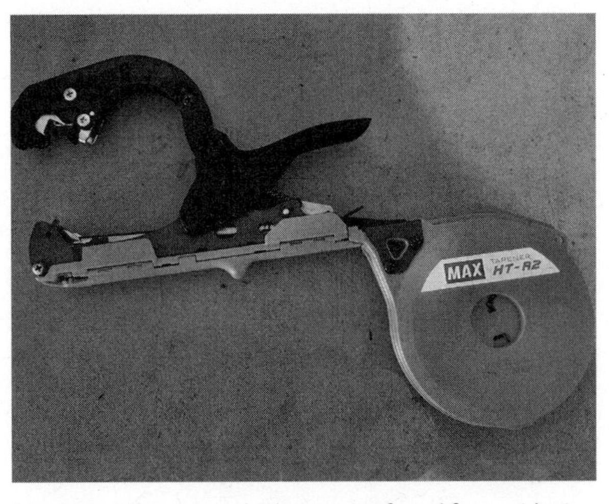

Max brand Tapener is the best we've found for rapid tying.

Berry Rake: This handheld scoop-like tool has fine comb-like protrusions on the scoop end that are used to comb through and rapidly harvest small berries like blueberries, currants, and gooseberries. Much faster and more efficient than hand picking. Different styles and prices are available online.

Credit: Johnny's Selected Seeds

A quality berry rake can make harvesting blueberries much faster.

Hoe: This classic farm tool varies drastically in quality. The very best ones are usually made in Europe and available through specialty tool suppliers. Good hoes are made of strong steel, are sharp, with solid wood handles and have properly angled blades for slicing down weeds efficiently. Hoop hoes (stirrup hoes, hula hoes) are also highly efficient and useful. These are a must for most small market growers, but invest in a good European one. Keep it sharpened.

Spade: When planting trees, you need a sharp, flat-bladed *spade* as opposed to a scoop-like shovel. Shovels are designed for moving loads from point A to point B, and are scoop-shaped to acquire and hold loads. Spades are sharp digging tools that make digging holes easier and faster. Keep them sharp and clean.

Harvest Containers: Go for the kind that are a tub that hangs on your belly and is strapped like a backpack around your shoulders. Then you can use *both hands* to simultaneously harvest, making harvesting much more efficient. This is probably the best way to go. Loads from the harvest tub can then be put into larger crates. You must not stack the fruit too high or it will crush, and it cannot be left in the sun. You might find

Three wooden quarts fit inside a Johnny's Harvest bucket. You can put fruit directly inside them and then it's ready for market. Or you can harvest into the bucket to sort later.

directly packing it in the field into quarts or clamshells to be a good way to go. A pull cart (or four-wheeler with cart) can be useful to move the crates around.

Collapsible Harvest Crates: Foldable, stackable, collapsible plastic produce crates are extremely handy. These are usually black or tan colored, and about 6" (15 cm) deep or so. While found online for upwards of $20 each plus shipping, they can be obtained used (once) for very cheap or sometimes for free from grocery stores, fruit markets, and fresh produce vendors, who sometimes have large stacks of used crates and are eager to sell them off, often for a couple of bucks each. Before buying, make sure the locking mechanisms on the short ends work properly and lock securely, so that they don't collapse in on themselves after being assembled and pressed upon. Make sure they are not cracked or damaged, or greasy. Wash thoroughly before reusing. Buy a few extra in case some

These repurposed produce crates are an asset on the produce farm. Note collapsed crates in the background.

break on you. I can get these locally for about $3 each. Search these out! Depending on the size of your operation, you may need dozens or even hundreds of these crates.

Coolbot®: This extremely clever and handy gadget has been a game-changer for many small growers, including us. Basically, it is an electronic device that "hacks" into a window air conditioner unit and allows the unit to cool down much colder than usual, down to 33°F (0.6°C). The retrofitted air conditioner unit can then be installed into a window or opening in a heavily insulated shed, existing walk-in fridge unit, or a retrofitted back room—effectively turning any retrofitted room or small building into a walk-in cooler.

Walk-in coolers are a must for most small fruit growers and micro farmers. However, they are very expensive, with very small ones starting around $10k. A Coolbot® retrofit can be done for a few thousand, including the air conditioner units, which go for about $1–2k brand new. For about $6000 we turned a 12 × 28' (3.7 × 7.6 m) insulated metal building into a huge walk-in cooler, utilizing a Coolbot® modified window air conditioner. It's recommended for a building this size to utilize two units, but we only run it December–April when the weather is cool and it does what we need.

9

Maintenance, Protection, and Organic Pest Control

Integrated Pest Management (IPM)

All fruit planting requires a certain degree of maintenance, protection, and pest control. This can all be done organically with proper knowledge, tools, and strategy.

First, understand the basics of Integrated Pest Management (IPM). IPM is a strategic, sensible approach to pest and disease management which began in the 1970s.[1] IPM focuses on key strategies:

Proper training and pruning

This varies by species and even sometimes by cultivar. Training and pruning accomplishes a number of crucial goals: lessening disease pressure, increasing quality and yield, making harvesting easier and more practical, allowing for netting by keeping plants smaller, increasing the strength of the plant itself (reducing breakage). Make sure to study the plants you intend to grow, both in the field and in university publications, and prune/train accordingly. *Appropriate cultivar selection* for disease/insect resistance is also crucial and foundational to IPM, and was discussed in depth earlier in this book.

Hygiene

Plantings and growing areas, including tunnels, must be kept clean and hygienic. Trimmings and pruned branches must be cleaned up. Dead,

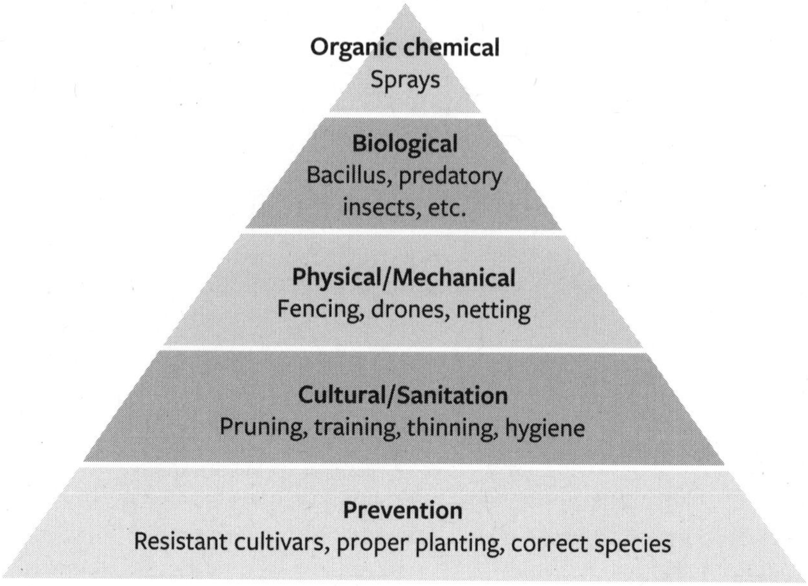

Organic chemical
Sprays

Biological
Bacillus, predatory insects, etc.

Physical/Mechanical
Fencing, drones, netting

Cultural/Sanitation
Pruning, training, thinning, hygiene

Prevention
Resistant cultivars, proper planting, correct species

infested, diseased, and dying plants and trees must be readily removed and burned. Keep the area cleaned, trimmed, and cleared around all plantings. Never allow prunings or dead plants to pile up, as this can create major insect problems. Keep high tunnels clean of rotting fruit and junk/debris. Also, keep horticultural tools sharp and clean by scrubbing and wiping with isopropyl alcohol. This is very important if you are pruning diseased branches or diseased plants—clean your snips/loppers in between plants.

Practice *biosecurity*, being careful when introducing plants from new nurseries and areas. Check for disease symptoms, insects, weeds (in potted stock or even entangled in bare rooted plants), and other issues before introducing them. Make sure to source plants and materials only from reputable, licensed nurseries. Inspect them carefully and diligently remove and destroy all weeds in potted specimens. Destroy virus-infected plants! Keep boots and shoes clean.

Without good hygiene, whether indoors or out, pest and disease issues will be exacerbated and much more problematic.

Organic approved supplements

We all want less work and less inputs in our agriculture systems. Inputs cost money, consume resources, and take time and labor to acquire and apply. However, if economics are important to your operation and you're not just a backyard grower, you have to account for the loss of plants or yields versus the costs of any treatments. Odds are, that $50 bottle of neem oil will pay back dividends in increased yields when used carefully. However, focus on training/pruning/hygiene and cultural practices (cultivar selection, proper planting, etc.) before reaching for a bottle.

Organic products continue to rapidly evolve and are more sophisticated, effective, and more specific for a wide range of disease/insect issues than ever before. The list of new organic products just keeps growing. Contact your local entomologists, ag extension office, and pesticide/ag chemical companies and inquire about organic options. Many effective pest controls can be made at home, including soap sprays, traps, sticky traps, compost and garlic infusions, etc.

Insect netting protecting celery crop in Germany. This can be useful for protection against stinkbugs and other insects on small plants like strawberries.

Physical barriers

These are very effective, but are only practical for insect control on small, shorter plants (strawberries, currants, etc.) which can be covered. Insect netting can be expensive and needs to be supported via hoops. Hoops can be made out of steel EMT tubing bent with a hoop bender. See Resources for sources for benders and also insect netting; EMT tubes can be purchased at many hardware and electrical supply stores. Although not ideal, it is useful to know insect netting exists and can be useful in some situations, draping this very fine netting over EMT tunnels can protect small fruit plants from flying insects. It is expensive and fragile. For deer and large animals, physical barriers are the only effective option.

Insect management

In addition to IPM, you can release and establish beneficial insects to help you manage problematic bugs. Mantises, lacewings, ladybugs, assassin bugs, beneficial nematodes, mealybug destroyers, the list goes on. Many of these and/or their eggs can be mail-ordered and safely released on your site. Provided with some flowering shrubs and shaggy areas of wildflowers, they should stick around and establish a population. They will help keep many problematic insects under control.

Unfortunately, due to the global trade of agriculture products, which mostly only benefits mega investors, mega farms, and major corporations, many highly invasive and detrimental insects (as well as many other species of plants and animals) continue to be regularly introduced to new areas, where they rapidly and permanently establish and cause irreversible and enormous economic losses, far in excess of the benefits of the temporary global trade that brought them in. There is a long list of such insects that is steadily growing. Fortunately, as has been observed, in due course many of these invasive insects are put in check by local predators, such as birds and wasps that adapt to feed on the new and abundant food source. Agricultural scientists also work hard to identify natural predators or pathogens and often can introduce them to bring back balance. This often works extremely well.

Credit: https://growingsmallfarms.ces.ncsu.edu/2020/04/birth-assassin-bug/

The large, tough *Arilus cristatus* is a voracious predator of many harmful insects. This predator is naturally found on our farm.

Some of the newest pests to invade the USA and elsewhere are below:

Brown marmorated stink bug

This new species of stink bug is extremely adaptable, mobile, and voracious, making it a big problem. In 2010, in the Mid Atlantic alone, over $37 million in apple crops were lost, and some stone fruit growers lost more than 90% of their crops.[2] *Halyomorpha halys* feeds on most types of fruits, berries, melons, and some vegetable and seed crops such as soybean and corn. It overwinters in large numbers in houses, siding, and other structures. Many other species of stink bugs, often native to North America, are also problematic and widespread. Research what species are in your area. *Control*: Still being researched. One of the best controls for brown marmorated stink bugs has been kaolin clay (Surround®). However, kaolin clay *cannot be used* on soft fruits like figs or berries, because it ruins those soft fruits. It works great on fruits with tough skin (apples, pears, etc.). Pheromone traps may help, as do pyrethrum sprays. Neem oil may possibly help deter them. On a small scale, stink bugs (and also Japanese beetles) can be hand-picked and dropped into a bucket of very soapy water.

Spotted lantern fly

Like the marmorated stink bug, the spotted lantern fly (*Lycorma delicatula*)[3] first appeared on US shores in Pennsylvania. It is a strange-looking flying insect native to Asia and India. It has spotted wings and a round, bright red body that becomes visible when it flies, hence the name. It is a sap-sucking insect that is currently a threat to grapes, stone fruits, apples and pears, and various hardwood and softwood species including pines and maples. It feeds in mass numbers through the trunk and branches of these species and this also causes sooty mold to grow in

Credit: Wikipedia Commons

the corresponding wounds, another harmful issue in itself.[4] It's a big deal. Controls are being investigated. Likely pyrethrum sprays and kaolin clay will be useful. The large congregations of the insects on the trunks of host trees is a strategic weak point, and could help with control measures, such as wrapping trunks in cardboard and applying sticky coatings, such as Tanglefoot.® Also, a quick pass with a propane weed torch across a mass of the insects could help. Effective homemade trap designs exist.[5] As of this writing it is only known to have established in PA but will be spreading elsewhere soon, and has been found in other states including Indiana, North Carolina, and New York. Be on the watch for this one!

Spotted Wing Drosophila and Fig Fruit Fly

Other newer invasives are the Spotted Wing Drosophila (*Drosophila suzukii*) hereon referred to as SWD, and the Fig Fruit Fly (*Zaprionus indianus*) hereon referred to as FFF. Both are similar "vinegar flies" or fruit flies drawn to very ripe or rotting fruit such as berries, tomatoes, figs, etc. They both infest soft ripe fruit (and even sometimes unripe fruit) by injecting eggs through the fruit flesh via their sharp ovidepositor (our native fruit flies lack a *sharp* ovidepositor and can only insert eggs into overly ripe or very soft fruit). Those eggs soon hatch and tiny maggots begin feeding and ruining the fruit.

Hygiene is crucial for management of SWD and FFF: keep all plantings free of rotting, damaged, and dropped fruit. Clean it up. Pigs and chickens can help with that. Also, keeping bushes such as red raspberries pruned properly to allow in sunlight will help, as direct sun desiccates the flies. It is for that reason they congregate in shaded areas, which should be minimized if possible.

Traps can be effective. For these, many designs and lures exist and can be found online.[6] Raspberry essence is apparently very effective as a lure. Refrigeration down to 34°F of freshly picked fruit deactivates and kills SWD larvae and eggs, saving the fruit and keeping it marketable.[7]

Credit: Wikipedia Commons

Spotted Wing Drosophila (SWD) fruit fly.

The fact that SWD reaches peak population only in late summer means that planting early and mid-season cultivars can help avoid infestation. Some growers, such as Guy Ames in NW Arkansas, say that the insect appears to be less of an issue than it was 5–10 years ago, as local fruit fly predators have adapted to feeding on SWD. The USDA is currently releasing parasitoid wasps that feed on eggs and larvae of SWD and FFF. Encouraging news![8]

Common animal pests

Deer

The only real strategy that works for deer is to erect physical barriers (usually 7–8' or 2.1–2.4 m tall deer fencing). As Michael Phillips encourages, *you'll come to love that big fence*. Fencing can be very expensive, or not, depending on the style of fence. Some find effective the lean-to fence[9] or a 3D fence.[10] Some make a deer fence out of tall, inexpensive, strong barrier plastic fencing like that used around construction sites, or thick, knotted plastic netting. The latter is actually a great option, especially if you're on a budget, as it is effective, very affordable, fairly durable, and as a plus is almost invisible from a distance. See the Resources section.

Tall, 10' (3 m) metal T-posts driven 2 feet (0.6m) deep are easier to install than wooden posts, sturdy, very long lasting, and cheaper than wood posts, and can hold either netting or plastic fencing. Whatever style,

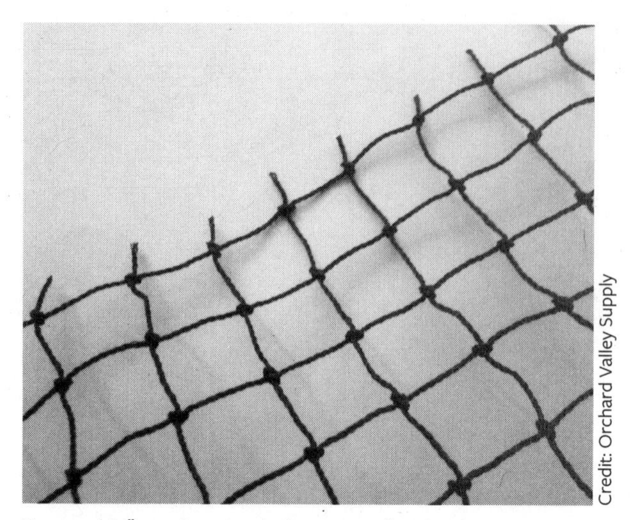
Credit: Orchard Valley Supply

Knotted ¾" mesh polyethylene rope fencing is strong, durable, and lasts 10+ years for a small fraction of the cost and installation work of metal fencing.

it must be sturdy and maintained—check for holes, breaks, and keep vines from overtaking it. Metal is obviously the best fencing material and lasts for decades. Deer will certainly test it, and are very strong and agile. If you let weeds, trees, and vines grow on and along the fence it will lessen its effectiveness.

Don't underestimate the capability of deer to penetrate into, and wreak quick havoc and destruction on an orchard or berry planting. It's inevitable to occur unless your site is well protected or you live in an urban or arid deer-free region. If the deer fence is lined along the bottom with tightly spaced wiring or chicken wire it can also help with *groundhogs, raccoons, and rabbits. For these smaller critters, relocation via live traps or guard dogs may be necessary.* Some people electrify their fences and apply peanut butter bait, which harmlessly shocks deer on the tongue, serving as psychological conditioning for the deer to *stay away.*[11] Cats can kill rabbits, keeping their population in check.

Birds

Birds are problematic, but unlike deer they only eat the fruit and not the plants themselves. However, their unchecked predation of small fruits can sometimes lead to 50–100% crop loss, making them a *very big deal* not to be underestimated. Fortunately, we have many very effective and modern tools in the dance with hungry birds.

First, remember that physical barriers are the best way to deal with encroachment by animals (or humans, or even insects). For birds, utilize bird netting, a thin, plastic netting draped loosely over berry bushes, vines, etc. Small growers sometimes also construct metal or wooden frameworks, sometimes quite extensive, covering entire plantings in bird netting. Make sure the netting has NO openings or areas where birds could enter, including going under a loose net. *They'll find it.* Also, if the net is pressed up against the fruit, the birds can often perch and peck at it with their beaks, defeating the purpose of the net. Make sure to re-cover plants as soon as you are done, or every time you move 25' (~8 m) down the row. If you forget to re-cover, birds can rapidly get in and pilfer the berries, or get trapped under the net.

Modern tools to consider include bird-scaring laser devices (said to be very effective) and noise making cannons (which temporarily work, and not suited to urban growers!). Rubber snakes, faux owls, and scare-eye balloons positioned around vines and bushes can help on a small scale, but must be moved around frequently to remain effective. Currently, flying drones are being tested as bird deterrents and this will likely evolve and become more useful.[12] Perhaps the drones could have attached a light weight, faux but realistic-looking predatory bird in flight. Drones can also emit recordings of distressed bird calls.

Credit: Wikipedia Commons

Voles, mice, squirrels, and rabbits

These are often an impactful problem. Voles are something like mini gophers or wild hamsters that have a strong appetite for the sweet bark of fruit trees and shrubs. Chicken wire or hardware cloth around the base of the plants is very effective. Cats and dogs are often very good at hunting voles, mice, gophers and rabbits and can really reduce their numbers. Gophers can be stopped from eating fruit tree roots by installing chicken wire baskets in to the planting hole at the time of planting the tree or shrub. Gophers especially like eating fig roots and can kill entire trees. Squirrels require live trapping, physical barriers, or hunting by cats or dogs. Their numbers must be kept in check, especially in urban settings where they have gotten out of control. Metal flashing sheets (available at hardware stores) can be tied around tree trunks (or nailed with small brads) and these provide an effective barrier to climbing squirrels, as long as they cannot jump into low branches or access trees via other structures such as fences, or nearby trees and plants. Attracting and housing owls would help too.

Raccoons, Badgers

Raccoons can be a serious challenge on "larger" small fruits such as figs, grapes, and tomatoes. Trapping with commonly available *large* live animal traps is the best control—these can be purchased at farm supply and many hardware stores. Also, some report the *Duke Dig-Proof Raccoon Trap* to be effective. Bait live traps with slices of watermelon or cantaloupe, cat food, peanut butter, sweet corn, or tomatoes. Release captured raccoons 5 miles away in a forested area or call animal control. *Badgers* can be an issue and may need to be relocated. Having a radio playing "talk radio" 24/7 can deter badgers and other animals. It deters me, also.

Humans

Humans can also be a serious challenge, as fruit theft is sometimes a real problem. For that a tall deer fence with a lockable gate does double duty. Urban settings are especially susceptible to human interference, including vandalization, theft, etc. Plan accordingly. Chain link fencing, security cameras and signage might prove useful.

For pest control, organic fruit growers have many potent and practical tools in the 21st century toolbox: pyrethrum, kaolin clay, neem oil, soap sprays, essential oils (clove, thyme, cayenne, garlic, etc.), *bacillus* and other beneficial microbial sprays, mail-order predatory nematodes and beneficial predatory insects, drones, laser devices, live traps, etc. These can usually be ordered right to your door. Just remember, these days for virtually every problem there is an effective organic solution out there—you just have to inquire and find it. Universities are a great place to contact if you are feeling confused or frustrated with certain pest or disease issues.

However, if pest problems (especially insects) are severe, check into these possibilities:

1. *Are you mono cropping or have too much of one crop?* Diversity may be needed in the planting, including preserving or encouraging flowering plants for predators, wildlife habitat for insectivorous birds, diversity of fruiting species, adding different cultivars, etc. Every organic farm needs shelter belts/hedges and wild border areas for beneficial wildlife and insects.
2. *Are your plants distressed or unhealthy and attracting insects?* Often insect infestations like aphids are a sign your plants are stressed. Are your plants not proper species or cultivars for your area? Are they water or nutrient stressed (including overly fed with nitrogen?) This can lead to infestations of various types.
3. *Are you encouraging pest-prey relationships through habitat protection and introducing predators (ladybugs, mantises, lacewings, cats, etc.) to the site?* Or are you spraying and killing everything, including beneficial predators? Are you killing or discouraging hawks, owls, and other animals that prey on voles, mice, gophers, squirrels, etc?
4. *Is your orchard hygiene, pruning, and weed control poor?* Is fruit falling and rotting under the plants, are dead fruit branches or plants lingering around, are your fruit plants getting overgrown and messy, are weeds out of control? This needs to be addressed with pruning, hygiene, and clean up. Burn diseased and pruned branches, don't let them pile up and rot—this can be a huge draw for problematic insects including *ambrosia beetle.*
5. *Is your planting site stressing out your plants due to being overly wet, overly shaded, overly dry/cold/windy, or in a frost pocket?* These stresses can exacerbate insect problems. Often if a single plant is infested, it means damage to the root system and the plant should be replaced if the problem persists. Overly wet sites can often only be made drier through installing drainage tiling. This is best ascertained and installed BEFORE planting and may not be possible afterwards. Your

local ag extension consultants can often come out (for free) and consult you on handling drainage issues.

Sometimes persistent agricultural problems are nature's way of saying "*this will not work here*" and you should listen and try something else. For individual pests of specific crops, research USDA databases, talk with your local agricultural extension office or your state entomologist, and consult with local growers.

Frost protection

In this era of erratic weather, late spring frosts and freezes can unexpectedly arrive. You need to keep up on local weather forecasts using a trustworthy app. You also need to be prepared with a few supplies and tools. Once you lose your whole crop to frosts or freezes you'll be looking for solutions. Some solutions are below:

1. *Frost blankets.* These lightweight white blankets come under a few trade names and brands. I like the thick heavier ones. They are more durable and can protect during colder temps. These are only necessary when small, sensitive plants are in bloom or setting fruit and frost threatens. Strawberries greatly benefit from being covered during bloom when a hard frost threatens. Many other small fruits (including blackberries, raspberries, gooseberries, etc.) are not usually affected by frosts during bloom, and are not able to be covered due to their size. Frost blankets are also good to have on hand to cover small newly planted berries in case of a late freeze. Keep them folded or rolled up on a stick of bamboo or similar and stored when not in use.
2. *Mist System.* All-night mist will coat plants and flowers in ice. This actually protects them from freeze damage, as long as the mist goes all night long, and the ice is dripping water all night long.[13] If done incorrectly, or if winds disturb the flow of the mist and dripping ceases, damage can result.

One evening in spring 2021 I rigged up a very fine misting system around our budding-out mulberry trees. It consisted of a $5 very fine mist emitter attached to a hose and strapped to a T-post via a screw-on pipe clamp. It misted the trees to the point of constant dripping all night long. By morning the small green shoots were covered in ice. When the sun had risen and air temps were warming, I turned it off. The shoots thawed out and were undamaged. This is worth researching and is a viable way to protect your crops, especially blueberries.[14]

3. *Orchard heaters/smudge pots/wind machines*. These energy-intensive methods have proved valuable for many fruit growers over the years. They are expensive and costly to operate, but can save the crop.[15] They are used in California and Florida on high value crops like citrus and avocados. You might research this further and see what might work for you. To protect a small planting, perhaps only a few heaters or fans might be needed.

Providing mist all night long during a freeze will cover shoots and buds in ice. If done correctly, they will thaw out once temps warm and be unharmed, as this mulberry shoot was.

PART 2

Getting to Know Your Fruit Allies

Many excellent small fruit options are available, but you need to focus on what is appropriate for your market, climate, and purpose. It's crucial to source locally adapted species and cultivars. To identify the best cultivars/species for your local area, research local university publications, consult with your local ag extension office as well as local growers and/or knowledgeable plant nurseries. Again, *never* guess on cultivar selection or base it on what sounds good in the pages of nursery catalogs.

10

Blackberries

The king of berries stands tall and strong on the micro farm, and easily puts out an encouragingly high yield of luscious, plump, juicy fruit when well grown. *Cultivars vary a lot these days and are very specific in terms of growth habit, ripening time, berry quality, etc., so choose very carefully.*

Blackberries thrive in hot and humid summer conditions. Certain cultivars and subspecies, especially raspberry/blackberry hybrids are best adapted to the cooler conditions of the PNW and coastal California, and some are adapted to low-chill areas and even subtropical conditions. Many species grow wild across the USA. In the Eastern USA most wild blackberries are small-fruited, fairly low quality, and seedy, but still harvested on a very small scale (and make great blackberry cobblers!). In the PNW many of the wild 'Himalayan' blackberries and wild hybrids are large and quite luscious.

Types of modern blackberries

First let's start with the *3 main types of blackberries*. They are differentiated by their starkly pronounced, varying growth habits. Here's some lingo to learn:

Trailing: These are the true "blackberry vines." They vary in vigor and size but the most vigorous can grow 9–12' (2.7–3.6 m) long or more in a single growing season. They are productive but need careful management and a stout trellis system because single vines and fruit load can weigh 50 lbs. or more. Not the best choice for the micro farm or small backyard, but can be utilized if necessary. These need lots of space, 8' (2.4 m) or more between plants, and careful pruning and training to keep them in bounds and productive.

Semi-Trailing: These have more vertical "spine" to them and make thicker, more sturdy canes that stand erect more like a raspberry—straight and tall. However, the upper half of the plants eventually vine somewhat, and the growth is very vigorous. 'Semi-Erect' seems to be a somewhat synonymous term for this type, but may also be its own category, with semi-erect cultivars having less trailing top portions and slightly less vigor. Otherwise not a lot of important differences apparently exist between the two, except that the University of Kentucky claims differences in yield estimates.[1]

Easier to manage and generally less vigorous than trailing blackberries, makes them preferable when the option is available. They require stout trellising. Modern breeding (except in the PNW) is focused primarily on developing semi-trailing/semi-erect or erect cultivars. They need about 6' (1.8 m) between plants. Semi-trailing plants are considered the highest yielding types, with 9000 lbs. per acre possible.[2]

Erect: These are the sturdiest and most shrub-like of all. The stems are rigid, thick and almost woody. They have medium vigor. *Erect blackberries are a great choice for the micro farm and backyard grower.* Trellis growing is best, but they can be adapted to non-trellis growing. For the backyard grower, large ringed metal tomato cages will suffice. They will not safely free-stand with no trellis; winds and storms will topple them over, or just the weight of the plant itself. These compact plants need only about 4–5' (1.2–1.5 m) between plants.

For market growers, *I recommend planting erect or semi-erect, thornless cultivars only.* They are more compact, easier to grow and support, and still produce excellent yields and berry quality.

That's not all, there's also:

Thorny: Self-explanatory; however, blackberries can be *very* thorny, with sharp, reticulated thorns. Most market growers are not going to want to grow thorny blackberries. However, if you live in a very heavy deer-pressure area and lack fencing, these can be capable of repelling most deer. The larger ones make excellent security hedges. And they often are *higher-yielding* than thornless blackberries. Many commercial growers still grow thorny blackberries for their high quality and excellent yields. Just wear gloves and be prepared to get pricked and scratched when pruning and harvesting.

Thornless: Zero thorns. What a horticultural achievement! In former days, thornless cultivars were considered inferior in taste and quality to thorny cultivars and they *were* inferior, being somewhat sour and seedy. That situation has changed with modern breeding in the last 25 years, and many newer cultivars are as good or better than the thorny cultivars, though perhaps not quite as productive. Much easier to manage, prune and harvest. Semi-thornless cultivars also exist.

Yet there's more. Blackberries also are divided into the two following fruiting habits:

Floricane: These blackberries produce flowers (and thus fruit) *only on year-old canes.* Meaning, year one: plants produce canes. Year two: those canes flower and fruit. This cycle continues, with every year new canes growing and maturing, and the following year they flower and fruit. (Then that cane dies and is replaced by the new ones growing around it, which will flower and fruit next season.) The mix of primocanes and floricanes assures production every year. Until recently this was the only type of blackberry available. *Pros*: Reliable, productive, and dependable harvests where adapted and by far most cultivars are floricane producers. *Cons*: Extreme winter weather and winter deer browsing can damage or destroy susceptible plantings. Pruning and training is required.

Primocane: Welcome to the new frontier in blackberries. Another horticultural achievement thanks to the University of Arkansas blackberry breeders. *Year one*: Canes emerge in spring, grow a certain height, around 4–5' (1.2–1.5 m), then flower and yield a crop. *Year Two*: If you pruned off the top growth that fruited in year one and allowed the canes to overwinter, you can harvest an early summer (floricane) crop, after which the cane is finished fruiting and dies. In late summer the new primocane crop ripens from canes that emerged in spring. Two crops possible per year. There are thorny and several thornless primocane cultivars available, with new ones currently being released by University of Arkansas.

Pros: Potentially no winter kill or deer browsing in winter: you just mow all the vines down after harvest. They will return in spring and yield a crop that same summer on the primocanes. Repeat. You will "only" get one harvest per season (late summer/early autumn) that way. *Or, you can overwinter the canes and have two picking seasons (spring and late summer) from both the primocanes and the floricanes.* The recent cultivars produce very large, tasty fruit. *Cons*: Hot, very dry summer conditions destroy primocane flowers, thus it diminishes or eliminates any harvest of primocane berries. This occurred with ours in the summer of 2019. Primocane blackberries do not set fruit properly in extreme or intense heat (90°+F, 32°+C). The mid–late summer flowering of the primocanes may prove an issue for that reason and also SWD flies can target late ripening berries. So, be careful when considering primocane blackberries if your late summers' temperatures are often 90°+F (32°+C) or SWD pressure is intense.

So, now if you read that a blackberry cultivar is semi-erect, thornless, and floricane producing, you should know exactly what that means.

Site considerations

Blackberries require full sun. Any shade drastically reduces yields. They do best on rich, clay loam soil such as good, well-drained pasture ground, but can succeed in many diverse environs and soil types including sand and even heavy clay. However, they require excellent water drainage and, for the safety of the blossoms, cold air drainage. They grow best in soil that's just slightly acidic and high in organic matter. Mulch as well as landscape fabric does a great job of weed suppression.

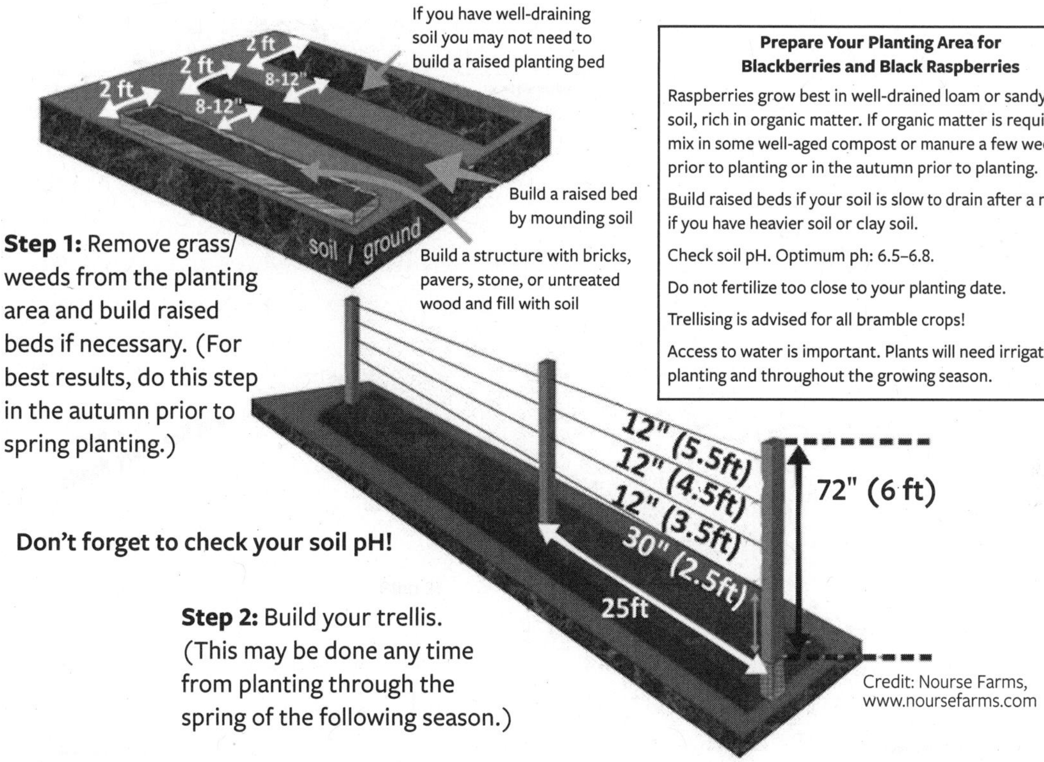

If you have well-draining soil you may not need to build a raised planting bed

2 ft
2 ft
2 ft
8-12"
8-12"

Build a raised bed by mounding soil

soil / ground

Build a structure with bricks, pavers, stone, or untreated wood and fill with soil

Step 1: Remove grass/weeds from the planting area and build raised beds if necessary. (For best results, do this step in the autumn prior to spring planting.)

Prepare Your Planting Area for Blackberries and Black Raspberries

Raspberries grow best in well-drained loam or sandy-loam soil, rich in organic matter. If organic matter is required, mix in some well-aged compost or manure a few weeks prior to planting or in the autumn prior to planting.

Build raised beds if your soil is slow to drain after a rain or if you have heavier soil or clay soil.

Check soil pH. Optimum ph: 6.5–6.8.

Do not fertilize too close to your planting date.

Trellising is advised for all bramble crops!

Access to water is important. Plants will need irrigation at planting and throughout the growing season.

12" (5.5ft)
12" (4.5ft)
12" (3.5ft)
30" (2.5ft)
72" (6 ft)
25ft

Don't forget to check your soil pH!

Step 2: Build your trellis. (This may be done any time from planting through the spring of the following season.)

Credit: Nourse Farms, www.noursefarms.com

Modern blackberries are generally super vigorous and need a lot of water and high fertility, so utilize nitrogen sources such as chicken manure, soy/legume products, fish emulsion, etc. A planting can stay healthy and productive for 8–10 years, sometimes more, after which it needs to be removed. Blackberries start production in one year from planting, wherein yields are about 25% of peak production; the peak is reached about 3–4 years from planting. A mature vine can produce anywhere from 3–20 lbs. or more of fruit, but a lot depends on the cultivar, environment, and training system utilized.

Cultivation

Blackberries can be established from rooted 1-year old plants (canes) as well as from viable root cuttings (certain cultivars only). They both give nearly identical results, with roots usually preferable and much cheaper to utilize when possible. Examples of cultivars that can be grown from root cuttings includes: 'Kiowa', 'Ouachita', 'Arapaho', 'Navajo', and 'Osage'. Please note that many blackberry cultivars are patented and propagating them without license may be illegal. Check into the current patent status on any blackberries before doing root cuttings or selling propagated plants.

Blackberries are pollinated readily by honeybees and other pollinators. Honeybees produce a high quality, light honey from blackberry and it's considered a premium browse. Two or more cultivars will ensure good cross pollination, which increases yields.

Starting with roots

If planting roots, ONLY use fresh, cold stratified (100 days at 34–40°F, 1–4°C) dormant root cuttings that have been stored properly and are fresh and free from mold and rot. They should be fleshy, about pencil-width, and not wrinkled or dried up. Tiny fibrous roots will not work.

Plant roots about 1–2 inches deep *horizontally* (like a fallen tree parallel with the ground, *not* vertically like a carrot root) in prepared, loose soil. Plant at the time of your last frost date or about a week or two before. They will sprout in about 2–4 weeks and growth first appears as tiny reddish-maroon leafy shoots. The new shoots can handle a light frost. If no growth is seen in 5–6 weeks and the weather is warm, something went wrong and you should dig some up and inspect, and probably replant ASAP, or utilize rooted plants at this point. You can prepare a planting bed with sileage tarps or tilling, and then flame weed it once or twice *after* planting the roots but well before the shoots emerge, say around 10 days after planting. That way you'll have a nice clean row with few weeds to contend with.

Planting dormant plants

If you are starting with rooted plants, plant *dormant bare-root canes in spring only* around your last frost date. Autumn planting is risky and plant loss could be very high in cold winter areas.

At planting, *cut off and remove any above-ground canes that are present...* What? That's right, cut 'em down! Any harvest you may get off the current foot-tall floricanes will be minimal and will set back the plant from establishing and getting off to a great start. It's *NOT* worth the setback for a few berries that will be a sub-par representation of the actual quality anyway.

Fertilizing

We feed each plant about 1 pound (0.45 kg) of 4-5-4 granulated chicken manure after growth starts in spring. If early summer growth is not vigorous and bright green, give more nitrogen and nutrients. Kelp extract can provide trace minerals. Newly emerged primocanes should reach 6' (1.8 m) tall or taller by June or July.

Mulch and irrigation

Blackberries need careful weed control while establishing. Straw makes an ideal weed controlling mulch. The University of Kentucky studies show that straw mulch increases yields more than using wood chips as mulch.[3] Mulch also conserves moisture.

Blackberries are drought hardy and once established do not require irrigation except in dry climates. Irrigation during initial fruit setting can increase yields but too much near ripening will water down the flavor and can oversize the berries. Utilize orchard tubing with 2 emitters per plant. Exact watering needs per season will depend on your region and soil type. Make sure they are receiving enough water—1 inch of rain per week or 2 inches of drip irrigation per week during the growing season. Do not use overhead irrigation on blackberries.

Pruning and training

Trellising is necessary for most cultivars. The end posts (6–8", 15–20 cm) in diameter) should have 6–6.5' (1.8–2.0 m) above ground and be well braced by a ground anchor. Place other posts 25–30' (7.6–9.1 m) apart in the row. These should be at least 4" (10 cm) in diameter and should extend 6 feet above the ground. Stretch two loosely stapled (10-gauge) wires at heights of 3' and 6' (0.9 m and 1.8 m) above the ground and securely fasten these to the end posts. Make it so that retightening each of the wires is possible each spring at one of the end posts via a ratcheting wire stretcher/tightener.[4] Utilize easy to use Gripple® anchors to anchor the end posts.

If you cannot or will not use treated pine posts, then black locust (ideally), osage orange, cedar/juniper, white oak, or other rot-resistant woods will work just fine. Trellis building is not especially easy but once you build a few you will understand it better and it gets easier. If you are not physically able to build them, consider hiring professional help. You may also consider planting only *erect cultivars* and using large tomato cages. Check the Resources section for sources of quality trellis materials.

Training: The following explanation is for training *floricane* blackberries and is identical to the annual pruning and maintenance process for floricane blackberries.

1. Plant the bare-root plants/roots as described, pruning down any canes present at planting. When the canes reach about 1–2' (30–60 cm) in height, *remove all but the best looking and strongest three canes. (You may only have one or two shoots the first season, leave them all if that's the case.)* These are the primocanes, or 1st year canes. Keep them straight

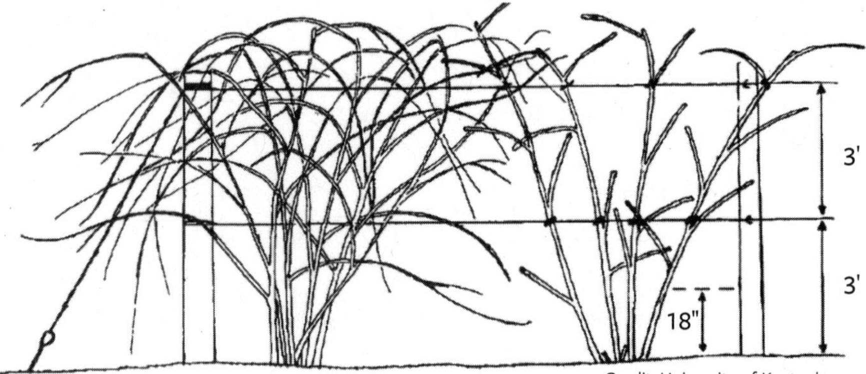

All growth below 18" (46 cm) is pruned off. Two wires for trellis are used, one at 3' (90 cm) and the next around 6' (180 cm).

Credit: University of Kentucky

and vertical by tying them to a bamboo cane (stuck in the ground at least a few inches away from the shoot), and tie or tape them to the trellis wires as they grow vertically so they are supported and do not fall over. Canes left to grow tall and not supported often snap over and die.

2. When those *three canes* reach the *desired height* (*usually 5–6', 1.5–1.8 m*), tip them (prune) so they start branching and stop growing vertically. Around mid-June or early July (earlier down South), *tip* them by removing the tender top six inches (15 cm) with your hands or snips. Make the height reasonable for your trellis, body height, etc. I do it at about 6' and the next year's resulting berries hang at around 3–6' (1–1.6 m).

   Soon after tipping, the cane will be stimulated to vigorously start branching just below the removed portion. Those branches are called *laterals* and are what will eventually flower and fruit next season.

3. As the canes grow, remove any branches that form below 18" (46 cm) high. Those low branches will be shaded and berries will be in the grass; remove them as soon as possible.

4. The following spring, last year's primocanes are now floricanes (one-year old flowering canes). Just as growth resumes (*not during winter*), prune the drooping floricane branches (*laterals*) to about 18" long. Remove any low branches at the bottom of the canes, below 18".

In spring, leave the best 3 canes, removing at ground level all small, spindly canes.

In early summer, tip the canes at the desired height. Branches (laterals) will soon form below.

Next spring (2nd year) prune all laterals to 18" and all low growth. Laterals will flower and fruit

Last summer, some laterals may have reached several feet long or even rooted into the ground. Cut them all to 18" long each. These pruned floricane branches will soon flower and set fruit this 2nd season. *Repeat steps 1–3 for the new primocanes that appear in spring.* When training and pruning, be careful as the fleshy branches and canes can snap off. After pruning, any rooted laterals you come across can be cut to 1–2' (30–60 cm), dug up and replanted, or sold.

I've found that circular, black plastic tomato support clips work great for trellising blackberries, and are surprisingly sturdy and reusable for several seasons. Trellising tapener devices work very well too and are fast. (I like the Max® tapener best.) See the Resources section.

Hygiene: Any time they appear, remove any dead, wilted or diseased/infested looking vines or shoots. Red-necked borer (insect) damaged canes will have large, cracked bulges on the stems. Best to prune off and burn any infested or diseased growth.

Pests/Diseases/Challenges

The biggest challenges with blackberries seem to revolve around a few insects highly drawn to them, often in large numbers. Japanese beetles, June beetles, red necked borers, SWD, and stinkbugs of various species being the most troublesome. SWD flies are mostly an issue on later-ripening cultivars. Deer can feed on the vines and foliage. See Chapter 9: Maintenance, Protection, and Organic Pest Control. Some specifics are below:

Pests

Japanese beetles: In our region, these and the June beetle populations seem to peak right when mid-season blackberries start ripening in earnest around the second week of July. Their feeding makes a portion of the berries unmarketable, perhaps 10–20%. Stinkbug feeding damages individual fruit drupelets turning them tan or white and unappealing, although the fruit is still often locally marketable (including as value added goods) and tasty (yet occasionally they leave a "stinkbug aroma"—gross!). Handpicking into buckets of soapy water might help but is labor intensive and unpleasant. Pyrethrums provide rapid knock down (killing) but *cannot* be used on ripe berries or during flowering.

Picking berries in the early morning and keeping the vines well-picked minimizes damage from Japanese and June beetles, which peak during the heat of the day.

Rednecked borers (RNB): The other, much worse insect challenge is RNB, *Agrilus ruficollis*.[5] This little elongated black beetle lays eggs on blackberry canes. The eggs soon hatch and the larvae burrow into the primocane vines, sometimes (often) *ruining that entire cane*, and up to 50% of the total canes in severe cases. Often the infested primocanes will appear fine and may overwinter and even bloom the next year, but then will suddenly and rapidly wilt and dry up due to the borer damage to the cane's vascular system. The tell-tale sign of RNB is a cane with a large swollen bulge with little cracks, often near the base but sometimes higher up. Guy Ames says to remove (and destroy) any infested canes as soon as you notice infestation. This may necessitate removal of a large portion of the canes in doing so, but will help manage the pest and reduce next year's population. Keeping wild brambles cut and away from the planting can help.

Using neem oil sprays on the canes a week before flowering and after petal fall might assist in insect control. Keeping the canes strong, healthy, vigorously growing and weed free also goes a long way. Urban growers should see less insect issues. Make sure to never spray pesticides or neem at flowering. Overall, blackberries are very tough and most problems don't seem to hold them back much (except for RNB infestations!).

Frosts and winter weather

In difficult, frosty spring weather, blackberries seem to nearly always prevail. We have seen many cultivars go through several severe late freeze and frost events in spring with fully developed flower buds on the canes. They continued to flower and fruit profusely. They handle extreme weather and late frosts very well, with flowers hardy to about 27°F (–2.7°C). They are adaptable and resilient plants well suited to the Southeastern US, up to zones 5–7 Midwest, West Coast, and warmer areas of New England.

In zones 4 and 5, source the hardiest cultivars available. Avoid areas exposed to extreme winter winds and plant windbreaks. Many growers in zone 4 *take the canes off the trellises in autumn and lay the canes flat on the ground, even covering with straw, to insulate.* This keeps the canes protected via ground heat and being out of the wind. Certainly you would need vining types in this regard as erect cultivars would snap in half at such treatment. See what types local or regional growers are utilizing and follow suit. *Note*: canes infested with RNB are prone to winter damage.

Finally, you can plant early, mid, mid–late, and late ripening cultivars if you have the demand for the berries all summer and want to stretch out the season. Late season berries may be affected by SWD. See Recommended cultivars below.

Harvest

Harvest the berries when dark black, shiny, and when they easily come off the calyx. Old-timers say they "should fall into your hand." However, that may be too ripe for market growers, so observe closely how they taste, pick, and handle. If they are too soft and leaky, you may be picking a little too ripe, or your cultivar may not be well suited to market. Shelf-life is short so chill as soon as possible and sell within 1–3 days. Some cultivars leak juice more than others (such as 'Triple Crown') so choose cultivars carefully. Harvesting in the chilly early morning hours pre-dawn will minimize beetles' access and damage to ripe berries, as they are much more active in the heat of the day. Modern University of Arkansas cultivars are bred with the commercial grower in mind, so should be some of the best suited to handling and not leaking or rapidly fermenting, etc.

Marketing

Pack in ½ pint to quart clamshells or quart berry boxes. Berries may leak if packed too tight or too high. Commercial packers use a little tissue in the bottom of the clamshell to soak up juice and this may prove useful. Some cultivars produce berries so big that only a few will fill up a ½ pint. Value-added goods are in demand and popular: jams, jelly, wine, cobblers, juice, ice cream, smoothies, fruit leather, syrups, and as flavoring for other products; there are many opportunities with blackberries. The dried leaves make excellent and healthful caffeine-free herbal tea similar in color and flavor to black tea. Plants are in high demand.

Recommended cultivars

There are a number of different species and also complex hybrids out there, but to keep matters simple we're going to just focus on cultivar names.

Ouachita: Developed by the University of Arkansas (UA), and considered by them their overall very best cultivar. The plants are adaptable and the berries are very prolific and especially sweet. Mid-season. Thornless. Semi-Erect. Zones 6–9.

Arapaho: Developed by UA, this early season berry is productive, easy to care for, very sweet, and berries are medium sized. One of the best erect cultivars. Thornless. Erect. Zones 6–9.

Kiowa: Highly productive, resilient, and *very* thorny deer-resistant cultivar that is early-ripening and has a long harvest season of about six weeks. Berries are extremely large and have very good to excellent quality. The berries are sometimes too large to fit in ½ pint clamshell containers. Very vigorous, adaptable, and resilient cultivar ideal for homesteading or processing, due to heavy yields and huge berry size. Zones 6–9.

Triple Crown: Extremely vigorous and needs stout trellising due to large size and heavy weight. Berries are abundant, sweet, and flavorful, ripening in mid-season. Up to 22 lbs. (10 kg) of fruit per vine has been claimed. Best suited for home and You-Picks, especially suited to the PNW. Some resistance to red-necked borers. Not as resilient or adapted in the south or Mid-Atlantic as newer UA releases, but remains popular. Thornless, semi-trailing. Zones 6–9, although extreme winters in zone 6 may lead to minor damage.

Chester: An older cultivar that performs *better than most in colder zone 5 areas.* Berries have very good flavor—sweet, yet slightly tart. Ripens a week or two after 'Triple Crown' and produces for about 6 weeks. Very productive, resilient, vigorous, and an excellent choice in colder regions and the upper South. Harvest season is very long, but berry size and quality steadily goes down as the season progresses. Excellent backyard or You-Pick variety. Thornless. Semi-Erect. Zones 5–9.

Hull: Developed by UA, 'Hull' is a good, thornless blackberry that makes sweet plump berries. Semi-Trailing yet fairly erect. Zones 6–9.

Illini Hardy: Developed by the University of Illinois for northern growers. Plants apparently reported to survive temperatures down to –23°F (–30°C), although is that roots or tops? Erect, hardy and vigorous plants produce good yields and high quality, slightly tart fruit. Ripens mid to late July. Thorny. Zones 5–8, maybe colder.
Sweet Ark® Caddo: A newer release developed by UA, 'Caddo is a reliable and heavy producer of superior large berries with high sugar content. Thornless, semi-erect vines.
Osage: Developed by UA, 'Osage' is a reliable and very heavy producer of superior large berries with high sugar content. Handles rain and cloudy weather better than other blackberries. Thornless, erect cultivar. Zones 6–9.
Darrow: An old heirloom cultivar that is very cold hardy and produces good quality berries. *Said to be hardy to zone 4*, but check with local growers or trial only. Thorny. Semi-Erect. Zones 5–9.
Sweet Ark® Ponca: Newest UA release, 'Ponca' has, according to them, achieved "the pinnacle of flavor." Very high sugar content and outstanding flavor. Productive, thornless canes. Quality remains consistent and high all season. *'Ponca' flowers in two successive waves about 10–14 days apart*, thus the plants have built in "frost insurance": in case a frost takes out the early flowers, it will still produce. Exceptional. Semi-Erect. Zones 6–9.
Prime Ark® Freedom: Developed by UA, this is the first-ever thornless primocane-producing blackberry. Blackberries are long and very large. Excellent sweet flavor, heavy production. Two crops per season possible (June and August). Extremely hot, dry late summer weather may destroy the primocane flowers, so be careful if this describes your local climate. Thornless. Semi-Erect. Zones 6–9.
Prime Ark® Horizon: Developed by UA and released in 2020. Primocane-fruiting with very high yield potential, particularly on floricanes. Berries are large and firm with good flavor and retain excellent firmness in storage, with limited leakage and decay. Ripens with 'Ouachita'. Vigorous grower is considered 'semi-thorny'. Zones 6–9.

Prime Ark® Traveler: Developed by UA, this cultivar is the second primocane-producing thornless blackberry. Blackberries are plump and large. Excellent flavor, heavy production. UA considers this the *first shippable primocane blackberry*, meaning it is firmer and holds up better to packing. Two crops per season possible (June and August). Again, extremely hot, dry weather may destroy the primocane flowers, so be careful. Thornless. Semi-Erect. Zones 6–9.

Columbia Series: Developed by the University of Oregon. There are a number of recent superior hybrids in this series. Large yields of high quality, very large delicious berries. For growing in the PNW only or testing in small plantings elsewhere.

Marion/Logan/Tay: Hybrids that produce large yields of high quality, delicious berries. For growing in the PNW, Europe, and California only or testing in small plantings elsewhere. *Usually these do not succeed or produce substantial yields in the Eastern USA.*

Summary

Excellent berries, strong production, relatively easy and fast to grow. Many varying types to choose from. Hardy and reliable. Market interest and value is fair–excellent depending on area. Easy and fast to grow but takes up a fair bit of space and usually requires trellises. Only takes 1 year to get yields, 3 years to full production. Primocane varieties bring in a great advantage for many growers by extending the berry season substantially where summers are not too hot. Quality continues to be rapidly improving through university breeding. Should be much more popular. For market growers I suggest focusing primarily on erect thornless cultivars, as they are so much easier to handle, take up much less space, yet are still quite productive. *Profit estimate:* around $25–50 per plant, possibly more.

Urban Market Farming Rating—2/5: May not be efficient on a profit per square foot basis compared to more profitable crops such as tomatoes, raspberries, etc, especially considering the trellising needs. However, demand is there and the price can be high. Focus on erect varieties and alternative ways to trellis (tomato cages, T-posts, etc.)

Rural Market Farming Rating—4/5: Easy to grow, and without space restrictions this could be an excellent niche crop to get into. Make sure quality cultivars are chosen and demand is there. Give out free samples to dispel any doubts about the flavor of your "tame" berries, as some rural folks may refer to them—disdainfully. Picking requirements and final market for the berries should also be carefully considered long before berries ripen. Popular and ideal for You-Picks; plant a range of cultivars for an extended picking season from June–September.

Home Recommendations: Plant erect cultivars, either bare-root or plugs (or potted is fine if you're only doing a few) and place large 5–6' (1.5–1.8 m) tomato cages over each plant. Plant 2–3 locally adapted cultivars for cross pollination and extended picking season. Keep mulched and watered the first year. Keep up on pruning and thinning of canes and annual spring fertilizing. If deer are an issue try thorny cultivars like 'Kiowa'. Experiment with primocane varieties unless summers are extremely hot in your area.

Identifying Quality Bare-Root Stock: Root systems should be large and healthy, usually quite fibrous and long and not dried up or rotting. Plugs also work fine, but are not as quick to establish as quality bare-root plants. Canes should be around pencil-thick or thicker, and smooth. You may see small pink shoots emerging from the roots—be very careful not to break these off as these are the new canes and are a sign the plants are needing to be planted ASAP. Remove any weeds or grass rhizomes entangled in the roots, if present, and cut the roots cleanly to 12–18" (30–46 cm) at planting.

11

Blueberries

Ever popular, delicious, and nutritious, blueberries are a dynamic success story of turning a wild fruit into a global market superstar.[1] Although popular, productive, and lucrative, due to very specific soil and cultivation requirements as well as 4–5 years to useful production, be careful when considering this crop.

Species

There are a number of different species and types of blueberries, including complex hybrids.[2]

Lowbush (*Vaccinium angustifolium*): Native to Northern areas and Mid-Atlantic areas. Bushes only attain 6–24" (15–60 cm) tall. Often wild harvested in parts of Canada, Maine, and Vermont where very large wild stands exist. Zones 4–6.

Half-High (*hybrid between lowbush and highbush*): Usually attain 2–4' tall. More productive than lowbush varieties. Zones 4–6.

Northern Highbush (*Vaccinium corymbosum*): Cold hardy and very productive. From Kentucky north into the Midwest, these are the blueberries grown and are excellent for market farming. Highly productive and excellent quality. Zones 5–7.

Credit: Wikipedia commons

Lowbush blueberry showing copious yield of berries. Easier said than done!

Southern Highbush (*Vaccinium corymbosum/darrowii/formosum*): From about Tennessee south these are the more common blueberries grown and are excellent for market farming. Highly productive and excellent quality. Not as cold hardy as Northern Highbush. Zones 7–10.

Rabbiteye (*Vaccinium virgatum/ashei*): Common in the Deep South and native to that region. Not as cold hardy as Northern Highbush. Excellent for market farming. Highly productive and excellent quality. Zones 7–10.

Cultivation

Having originated in quite acidic, well-drained, high organic matter regions, they have very specific needs for soil pH, drainage, and soil type. Blueberries generally do best in freely draining sandy soils. They can succeed in loamy soils too, but clay soils are often out of the question unless very heavily amended, and doing this beyond a small row or two is usually impractical. Altering soil pH is playing with complex chemistry and can turn wrong very fast if you're not very specific and mindful. Heavily alkaline or chalk soils will not work. So, all that being said, I would not recommend most people try to get into blueberry production unless you are sure blueberries are already being grown profitably in your area, and you have very thoroughly done your homework on your site and soil type and are sure they are both suitable for blueberry culture.

A few technical things to go over: blueberry soil must be high in organic matter and have a pH in the range of 4–5.5. If it's over 6 pH, blueberries will not be able to assimilate iron in the soil and will express iron chlorosis symptoms. Too low of soil pH can result in aluminum or manganese toxicity issues. Soil organic matter should be over 3%. CEC (cation exchange capacity) should be *low*, below 12, as on sandy soils. Otherwise, changing the pH will prove unrealistic. Elemental sulphur can be used to lower the pH for blueberries. Make sure to get a soil test and consult your local ag extension office.

Great regions conducive for easily growing blueberries (species differ) include parts of Maine, Michigan, and New England (lowbush, half-high, and Northern highbush species), and the Deep South in sandier parts of Central and Southern Mississippi/Alabama/Georgia, Louisiana, and Florida (Rabbiteye and Southern highbush species). In the sandy soils of the Deep South and Gulf areas blueberries grow prolifically and yield huge, high-quality crops. For example, south-central Mississippi is where the "blueberry capital" of Biloxi is located, a major commercial blueberry area. California is also a major producer and they grow quite well in the PNW. Chile is a major producer in South America.

In Midwestern and Mid-Atlantic areas, blueberries can succeed if planted in very well-draining loamy or highly sandy soil that is heavily and scientifically amended and great care is given. I have personally planted them in KY on at least three different sites and they have failed every single time, despite carefully amending the loamy clay soil, careful selection of cultivars, good after care, etc. (yet I am trying again in 2022!) Another local KY farm grew blueberries successfully for years until they all suddenly died due to a deadly pathogenic soil fungus called *phytophthora* root rot, caused by excess soil moisture. Another local aspiring blueberry grower consulted with the local agricultural extension office about amending his soil prior to planting a hundred or more blueberries. He followed their recommendation exactly, amending the soil to adjust the pH to make it more acidic. He planted, and all the plants rapidly died.

Blueberries are sensitive, relatively slow growing plants and *generally not that easy to grow*, especially on any scale, unless, once again, you are in blueberry country or very dedicated to doing very careful and specific soil amendment.

Planting

Plant blueberries in either fall or spring. In Northern areas plant in spring only, to avoid freeze-thaw cycles that can heave young plants right out of the ground, as well as to avoid winter deer browsing and severe low temperatures. In the South, fall or spring planting is fine. Always heavily mulch new plantings.

If your soil is not mostly sand then it likely will require considerable amending. You should do this at planting or before, as it's not possible after planting! Replace ½ the original soil out of each 24" deep planting hole with *moistened* peat moss, mixed thoroughly with the native soil. A small amount of finished compost can be mixed into the peat moss which increases organic soil matter. Make sure the amended soil is firmly packed into the hole (but not compacted down) so the blueberry will not sink down after planting. A deeper hole is better for the shallow, fine root system, allowing it to stay moist and cool when summer gets hot and dry.

It is also recommended[3] at planting time to break up the rootball, especially on larger potted stock. Otherwise, the plants can have difficulty with extending their tiny roots out into the surrounding soil, thus leading to serious problems with nutrient and water uptake that can lead to failure.

After planting each bush, top dress with organic mixed fertilizers, water, and 6–8 inches of organic mulch. Raised beds bordered with untreated wood can hold the heavily amended soil and mulch, improving drainage, and also help with weed control. If the organic matter content of your soil is low, you should cover crop with grasses (sorghum-sudan in summer, rye/legumes in fall-winter) for a year before establishing any sizeable blueberry planting.

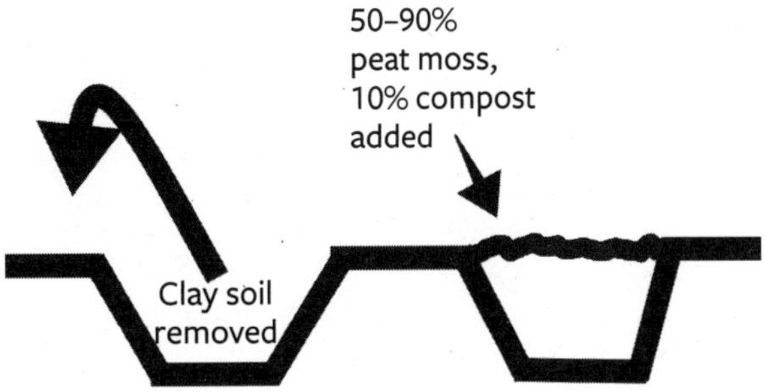

Blueberries are spaced within the rows according to species, with lowbush needing 1–2' between plants[4] and highbush 4–6' (usually 6' is recommended). Space rows according to your pathway maintenance equipment/mowers or lack thereof, and the height of the crop.

The plants need about 4–5 years of establishment before you get into substantial production, and are considered mature at age 7. Generally, 2-year-old bushes are what growers plant, often transplanted from 1–2 gallon pots. Always go with container-grown blueberries and not bare-root. Gently scratch the sides of the root mass to loosen the tiny feeder roots at planting. Myccorhizal innoculants are available specifically for blueberries, manufactured in Europe.

Very important to note is the recommendation *to remove any and all flowers and tiny fruitlets from the plants for the first 2–3 years.* Failing to do so may get you a tiny harvest of berries the first couple of years, but can *permanently stunt the bushes, ruining future yields.* So, you must be patient and strictly give the bushes 2–3 years to establish *before* allowing them to fruit at all. Some claim this isn't necessary, but better to err on the side of caution.

Blueberry cultivars vary rather dramatically, from bloom time to ripening date, size and flavor, so choose carefully. Some are more sweet, others more tart. Tart berries are excellent for baking and retain their true blueberry flavor in the oven, but may prove unpopular at the market. Some make large berries the size of nickels or even quarters, others smaller than peas. Some, like Rabbiteye types, are better in the hot South. Some take time to sweeten up past the initial blue color change, remaining tart for a while, such as 'Bluecrop'. There is a market for blueberry twigs and foliage for cut flower arrangements. For that, the cultivar 'Misty' makes especially lush and green foliage that persists into winter. Make sure your cultivar choices are fully compatible with your region, local climate and marketing route. Early-ripening cultivars also bloom earlier, and are thus more vulnerable to late freezes and frosts.

Weed control

Another challenge is *organic weed control for blueberries.* Many growers rely on using landscape fabric within the rows/beds. With sizeable holes burned or cut every spot where a blueberry will be planted, it provides excellent weed and grass control. The holes must be big enough to facilitate the natural suckering and shoot renewal of the bushes.

However, if possible, on non-weed infested sites, heavy organic mulch does a great job and provides acidity and abundant organic matter that blueberries need. Organic mulch options include pine bark, rotted pine wood chips, heavily rotted sawdust, chopped corn cobs, straw, and pine needles. It's recommended by the University of Kentucky to apply mulch 6–8 inches thick and reapply when it has degraded to about 3 inches thick.[5] On a small scale, cardboard, grass clippings, and fallen leaves also can work, but degrade rapidly in warm climates. You'll have to decide what works best for you. Don't plant blueberries on sites infested with weeds like pokeweed, bindweed and morning glory or rhizomatous weeds like Johnson grass, quack grass, etc. Eliminate those through intensive cover cropping and/or tarping first. Some amount of hand weeding, (sometimes a lot) is usually needed with blueberries, so consider that.

Although a great fruit crop, I wouldn't strongly recommend blueberries for beginner or even intermediate growers unless you're in a prime blueberry region and can dedicate to caring for the bushes properly. If blueberries are successfully being grown in your local area, try to inquire with the growers how they planted them, take note of the site conditions, and try to get cultivar and species names.

Irrigation

Blueberries are very sensitive to drought and require irrigation during the first few years of establishment, and in drier areas all season long. It's safest to have an irrigation setup when growing blueberries so drought will not kill the entire planting, or lead to laborious hand watering. Irrigation during initial fruit setting can increase yields but too much will water down the flavor. Overhead irrigation is fine as long as the fruit is not ripening. Stop irrigation as soon as fruit starts coloring up. Utilize orchard tubing with 3–4 emitters per plant. Exact watering needs per season will depend on your species, region, and soil type.

Pruning

The goal is to constantly renew the bushes by pruning out older and less productive canes, which often look very twiggy, and also are quite thick at the base. Encourage new ones to replace them through adequate fertilizing and protecting the bushes from deer browsing and string trimmers. Older canes will produce smaller berries. Canes are at their peak between 3 and 6 years old. Pruning is done only in late winter or early spring to avoid winter damage.

Old, thick, and twiggy growth and rubbing branches removed. Young branches and renewal shoots left.

Mature plants should produce 3–5 new canes per year, with ideally around 15 canes total per plant. Make sure old shoots are routinely removed at ground level and new ones are encouraged and allowed to bear for about 5–7 years. If no renewal shoots are allowed to establish, or accidentally eliminated, the bushes will decline in production and eventually die out.

Pests/Diseases/Challenges

Generally, if the soil and climatic conditions are favorable and good care is taken to grow the correct species/cultivars, problems are few. Most issues revolve around deer, weed pressure, and a few troublesome insects. Ag extension publications can name dozens, but mostly those are problems on large monoculture operations.

Deer: Large cages or fencing are the only cure. If you are putting in a sizable blueberry planting then consider deer fencing a necessity. Blueberry twigs are premium winter browse and will draw deer, as will the fresh foliage.

Mummy Berry: A problematic fungal disease. Keep bushes clean of infected, white berries and plant disease-resistant cultivars. Make sure to prune the bushes properly so they get good air flow. Mulch every fall to provide organic matter that buries any infected, fallen berries and prevents sporulation.

Cankers: These appear as red or purple spots on canes. Prune off and destroy during winter pruning if you notice any. Prune back to healthy growth.

SWD: Fruit flies such as SWD can infest late-season berries. Best to plant early and mid-season cultivars.

Caterpillars: Certain regions, including the Mid-Atlantic, can host various foliage-eating caterpillars. Growers report these can defoliate entire bushes in a day. The foliage grows back, but watch out for these. Bt products (*Bacillus thuringiensis*) and wasps may help.

June bugs and Japanese beetles: Feed on foliage and can be issues; pyrethrum (post bloom) or handpicking can be utilized.

Birds: Blueberries are a prime food for many bird species and bird netting is *absolutely necessary in most areas*, even urban ones. You may bend EMT tubing to form arches over the planting so netting is easier, and keep bushes pruned to a reasonable height. Netting laid across the bushes is less effective and will often damage the bushes at removal. See Resources for pipe bender sources.

Late freezes and frosts: Blueberry flowers are vulnerable to being destroyed by late frosts and freezes at the time of bloom or even just after. Therefore, choose cultivars known to succeed in your area and handle local conditions. Lowbush cultivars can be covered in thick row cover to get through a frosty night. Larger bushes and larger plantings should be equipped with mist systems that can freeze and protect the blossoms[6] Early-ripening blueberries will be more prone to damage as they usually bloom earlier. Blueberry blooms can withstand 28°F (–2°C) without damage.

Cultivars (by region)[7]

New England

Northern Highbush:

- Bluecrop
- Blueray
- Patriot
- Nelson
- Jersey
- Northland (Half-High)

Mid-Atlantic and Midwest

Northern Highbush:

- Atlantic
- Berkeley
- Bluecrop (standard commercial cultivar)
- Bluejay
- Bluetta
- Patriot
- Blueray

- Chandler
- Coville
- Duke
- Elliot
- Lateblue
- Sierra
- Toro
- Reka (exceptional flavor)
- Northblue (Half-High)
- Northcountry (Half-High)
- St. Cloud (Half-High)
- Northsky (Half-High)

Southeast

Southern Highbush:

- Gulf Coast
- Sharpblue
- Biloxi

Southern Rabbiteye:

- Tifblue
- Climax
- Powderblue
- Premier
- Windy
- Beckyblue
- Bluegem
- Bonita
- Brightwell
- Snowflake
- Woodard

Pacific Northwest

Northern Highbush:

- Toro
- Reka
- Blueray
- Bluecrop
- Earliblue
- Jersey
- Spartan
- Patriot
- Liberty
- Darrow
- Powderblue (Rabbiteye)
- Tifblue (Rabbiteye)

Harvesting and marketing

Mature blueberry bushes can produce anywhere from 5–20 pounds of fruit and can be tedious to harvest. The yields vary based on species and growing conditions, with highbush types producing the most fruit per bush. One large bush can take 15–30 minutes or more to harvest ripe berries by hand, and will need to be harvested multiple times throughout the season. Comb-toothed, blueberry *harvest scoops* exist that may prove very useful—see the Resources section.

Each cultivar of blueberry will produce ripe fruit for about 2–3 weeks. By having cultivars that ripen at different times you can stagger and extend the harvest through most of the summer and into autumn. Note: late ripening blueberries will have the potential for strong SWD infestations. Blueberries ripen about 60–80 days after flowering occurs.

Labor will increase steadily as the number of bushes increases. Make sure to factor in labor needs, which mostly revolves around a few summer weedings, renewing mulch 1–2 times a year, winter pruning, and harvesting the berries. Figure about 1–1.5 hours per year per bush all things considered. Thankfully, they are productive enough that even adding just 20–40 well-grown bushes to an existing small operation could prove quite lucrative at the farmer's market. The berries hold well in cold storage for about 5–7 days, so picking can be done throughout the week. Some cultivars have a "dry scar" which holds better in storage than cultivars that have a "wet scar" when picked (tiny opening in the flesh from tearing off the stem.)

When packing, ½ pint (237 ml) clamshells and pints work well. Wholesale prices for organic ½ pints are currently about $3.50–4.50, retail about $5–6. Get your market plan in order years before the berries will ripen. Hype is easy to build with this ever-popular fruit.

Summary

Rewarding yet challenging to grow except in certain conducive regions. Fruit is perpetually in demand and price remains high. In conducive regions this is an excellent crop and fairly easy to grow. Labor intensive to harvest and keep weeded. You'll wait 4–5 years before useful production and 7 years to full crops. Deeply research your area and soil type before proceeding. *Profit Potential*: Around $75–150 per mature bush, maybe more.

Urban Market Farming Rating—3/5: Takes years to get into production and many cities will not have proper soil requirements. Also, the bushes take up a fair amount of space. But if space and time are not important considerations, blueberries could be worth doing if soil type is right. If you filled a decent-sized urban backyard with, say, 30 healthy blueberry bushes, you could potentially earn up to $4,500 a year at full production, selling at premium retail prices.

Rural Market Farming Rating—4/5: Challenging to grow yet very rewarding where they grow easily, could be a 5/5 in good blueberry regions. Labor intensive to pick, so do not over plant and consider picking labor and final fruit destination. Good for organic production. A solid weed control plan must be in place, as well as effective deer and bird protection, and reliable irrigation. Popular for You-Pick operations.

Home Recommendations: Plant 1–2 gallon potted plants of at least 2 locally adapted cultivars in raised beds, heavily amended, preferably with peat moss or deeply rotted sawdust and compost—about 1 bucket of peat moss per blueberry plant. Apply sulfur before planting to adjust the ph as per a soil test. Keep mulched and watered all summer the first year. Bird protection is usually necessary at fruiting.

Identifying Quality Stock: Potted or container-grown is preferable. The bigger the plant the better, up to about 2–3' (60–90 cm) tall. Roots should be prolific, with many tiny feeder roots, and the roots should be firmly cohesive and not broken up. Brightly colored, often reddish stems, especially tall straight ones, are a sign of vigorous, healthy stock. Starts should not look bonsai-like or be overly branched.

Blueberry Soil Prescription

Lee Reich, PhD (www.leereich.com)

At planting:

Check drainage; plant on mound if drainage is poor.

1. Check pH; add sulfur if necessary to bring pH to 4–5.5. Use "pelletized sulfur". Sandy soils: ¾# per 100 sq. ft. for each pH unit. Loams: 2# per 100 sq. ft. for each unit.
2. Mix bucketful of peat moss with soil in planting hole.
3. After planting, mulch with 2–3 inches of loose organic material.
4. Water needs throughout growing season: ½ gal. per square foot spread, which is equivalent to 1" depth of rain or sprinkling per week.

Ongoing:

1. Test pH every couple of years or less to check that acidity is maintained; another indication of insufficient acidity is interveinal chlorosis of oldest leaves first; use pelletized sulfur again, if necessary.
2. Maintain a permanent mulch of wood chips, wood shavings, straw, or other weed-free organic material 2–3" deep, applied each year after leaf drop on top of old mulch.
3. Water regularly for at least the first two years.
4. Feed each fall before adding mulch: soybean meal at 2# per 100 sq. ft., or the equivalent amount of nitrogen from some other rich source, such as alfalfa meal (6# per 100 sq. ft. needed).

1# is 1 lb. = 454 g   1 US gal. = 3.79 L   100 sq. ft. = 9.3 m^2

12

Raspberries

The queen of berries bestows her fragrant, ruby-like tender fruits gracefully in great abundance. Red raspberries' ease of growth, small square footage requirements, strong production, and very high market demand make raspberries *one of my top picks for the micro, urban, and market grower.*

I'm surprised more growers don't grow this small, profitable, and ideal fruit crop. They require little space and start producing fruit in about six months from planting, even from root cuttings. With both June and fall bearing cultivars, the fruit ripens for an extended period as well. Let's explore this wonderful, feminine fruit.

Species

First, raspberries come in a number of species but the two most important ones you need to know are *Rubus idaeus* (red raspberry) and *Rubus occidentalis* (black raspberry.) There is some confusion out there because some people call black raspberries "blackberries" but they are not. Sure, they're both black in color and berries, but they're different species. All raspberries produce a hollow "thimble" shaped fruit, while blackberries have a soft but solid core.

Black Raspberries

Black raspberries are delicious and easy to grow, with their own unique flavor nothing like the reds; winey, tangy, and sweet. They are one of the highest in antioxidant small fruits, beating blueberries by 300%.[1]

Good selling point right there. Their main setbacks are a short season and modest yields. Primocane black raspberries have changed that, as we will look into.

The plants have small thorns and the season is early (late May–early July depending on location) yet the season is also short, about 2–3 weeks. Thus, they are not the best choice for market farming due to mediocre yields and a short season compared to other fruits, but can be grown profitably if the demand is high, space is not at a premium, and they still seem worth it. They yield somewhere around a ½ pint–1 pint (237–473 ml) per foot of row. So, a hundred-foot row and $5 per ½ pint could bring in about $500–1000 with good conditions and a viable direct retail market that can move lots of these berries fast.

The fruit is ripe when dark purple-black and separates easily from the calyx. It is soft, fragile and perishable but will keep 2–3 days in refrigeration if harvested when dry. Moist berries ferment in 1–2 days. Sell promptly in ½ pint and pint clamshells.

Insect pressure is low due to the early season ripening. Although early ripening, *the flower buds are frost tolerant and rarely get destroyed by late freezes, making them quite resilient.* Anthracnose on the canes can be an issue in very rainy climates and susceptible cultivars.

A well-maintained patch lasts about 8–10 years before rapidly declining. They require no trellising. In fact, many growers tip pinch the canes at 3–4' (0.9–1.2 m) to create a productive, easy to harvest, short hedgerow. Pruning is the same as for blackberries (except they are often kept much shorter). Pruning the tops and laterals makes the plants bushier, productive, and prevents them from tip-layering and spreading. They fruit on floricanes (1 year old canes) and right after the floricanes fruit they are finished and removed at the base.

However, recent primocane types have been developed that yield on same-year shoots (primocanes) and thus yield two crops per year *if primocanes are not tip pruned.* This makes black raspberries much more productive and potentially profitable to grow, but it needs more trialing to confirm. In our trials, 'Niwot' produced about 2× as much as all others (spring and late summer crops combined). The second crop ripens throughout August and into October. It's worth looking into and could be a much better option than traditional floricane producers.

I would only recommend planting black raspberries for marketing in an area where they are locally popular, otherwise they will not likely be worth it.

Red Raspberries

Red raspberries are my overall top small fruit recommendation for anyone wanting to get into small fruit marketing. Why? They are typically the easiest high value fruit you can grow (unless your regional climate is too hot and does not allow it). They crop in only six months to one year from planting. They are universally popular, easy to maintain, hardy and resistant to insects and disease, and need minimal space and maintenance. The fruit is strongly in demand and gets a high price (and has for over a hundred years). The fruit is beautiful, nutritious, easily sells, and is easy to harvest and process. It's possible to have ripe fruit for months at a time in most locations.

Two Main Varieties

The two main varieties of red raspberries are June bearing and fall bearing. This describes their harvest periods. June bearers flower and fruit throughout June and into July in most areas. Fall bearers are also called "everbearing" and these flower and fruit in August and continue until frosts and cold weather stop production. Fall bearing raspberries fruit on primocanes (first season canes) and June bearers fruit on floricanes (one year old canes) and so their maintenance is slightly different.

It should be noted that fall bearing red raspberries can also produce an early summer crop of berries on their floricanes (hence the name "everbearing"). However, this is *not recommended because it will lower the quality, yield and potential of the fall crop.* If you're aiming for a summer crop, plant June bearers. Don't try to double crop fall bearers; you'll simply get lower yields and lower quantity both times.

I recommend for marketing purposes, growing both varieties (June and fall bearers) simultaneously, as it can provide a total of about 3–4 months of marketable raspberries. Cultivars vary slightly in their ripening sequence and timing.

Cultivation

It is important to note that *red raspberries are generally not acclimated to southern areas with very hot and humid summers and usually will not survive. There are some newer cultivars such as Kweli® that can tolerate hotter,*

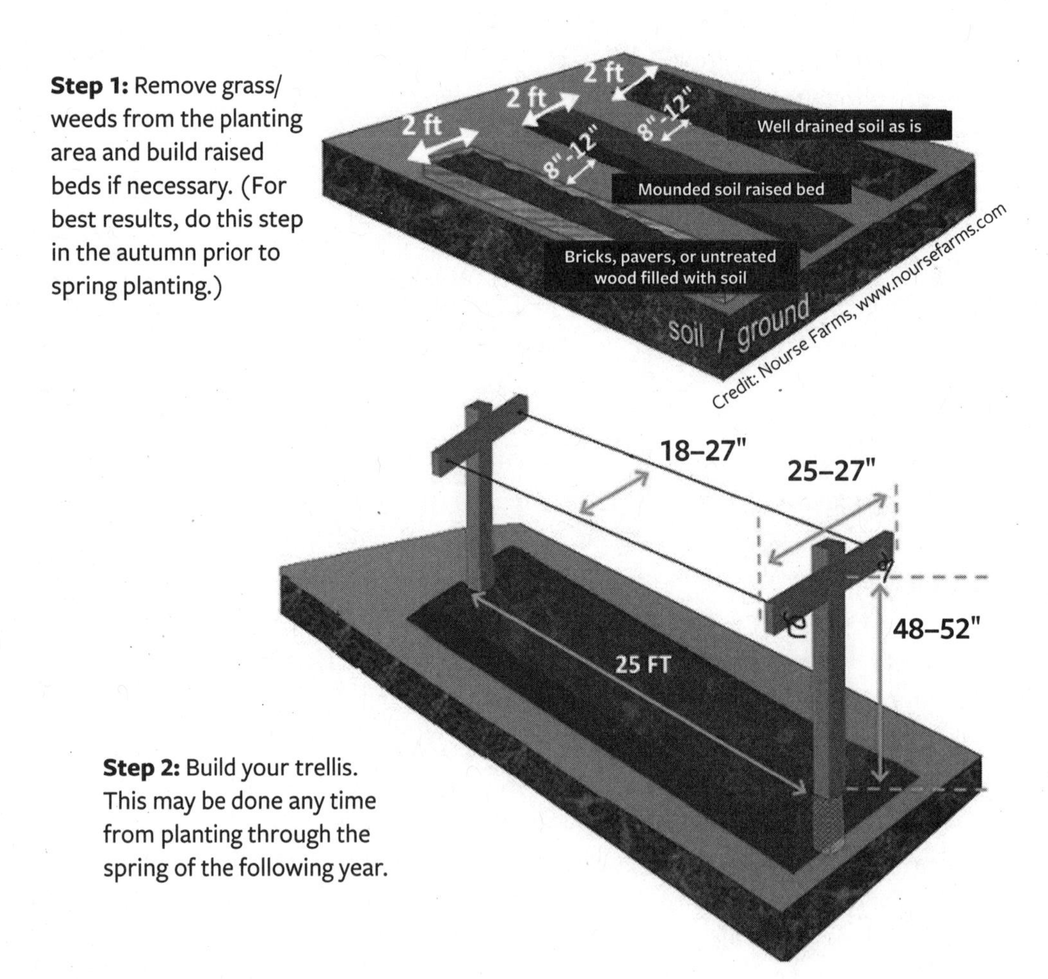

Step 1: Remove grass/weeds from the planting area and build raised beds if necessary. (For best results, do this step in the autumn prior to spring planting.)

Step 2: Build your trellis. This may be done any time from planting through the spring of the following year.

yet not humid, conditions. Virginia/Tennessee and mountainous upper south locations are likely their cut-off point in the south. Also, hot southwestern areas will not be conducive to raspberry culture (except with special cultivars as mentioned).

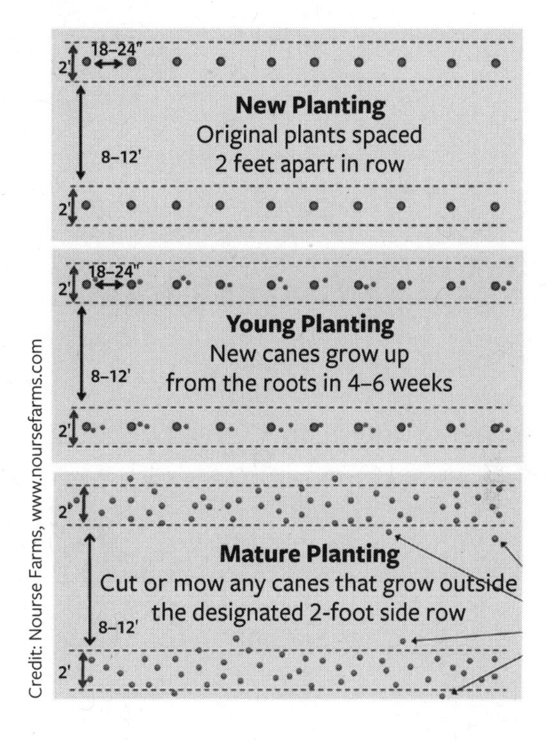

Credit: Nourse Farms, www.noursefarms.com

Red raspberries are markedly cold hardy and many cultivars can perform in zones 3–4. In very cold marginal areas you may want to focus on primocane fall bearing varieties and mow down after harvest. Black raspberries are slightly more tolerant of heat, and less cold hardy. Check with local growers and your ag extension office.

Growing beds for red raspberries should be 18–24" wide (46–61 cm). Too wide of a bed and the canes will become overcrowded and the berries will get shaded and difficult to harvest. Slightly raised beds work well as raspberries need very good drainage to avoid root rot issues.

The June crop sets on the floricanes (year-old canes) of June bearers. The fall crop sets on the primocanes (new canes) of fall bearers.

Do not mix up the two varieties (June and fall bearers), but grow them in distinct rows so you know how to manage each different type. Trellising needs are the same for both. For both types, maintain about an 8–12" (20–30 cm) distance between canes when pruning and thinning. Don't let them get overcrowded or production and plant health will decline. For June bearers, thin excess canes from the beds in winter or early spring, and for both types, thin excess shoots as they sprout and grow in early to mid-summer. Remember, for fall-bearers you are cutting them all down to the ground in winter.

When fertilizing, apply a general NPK organic fertilizer in early spring around the time primocanes start to appear, and again in May–June.

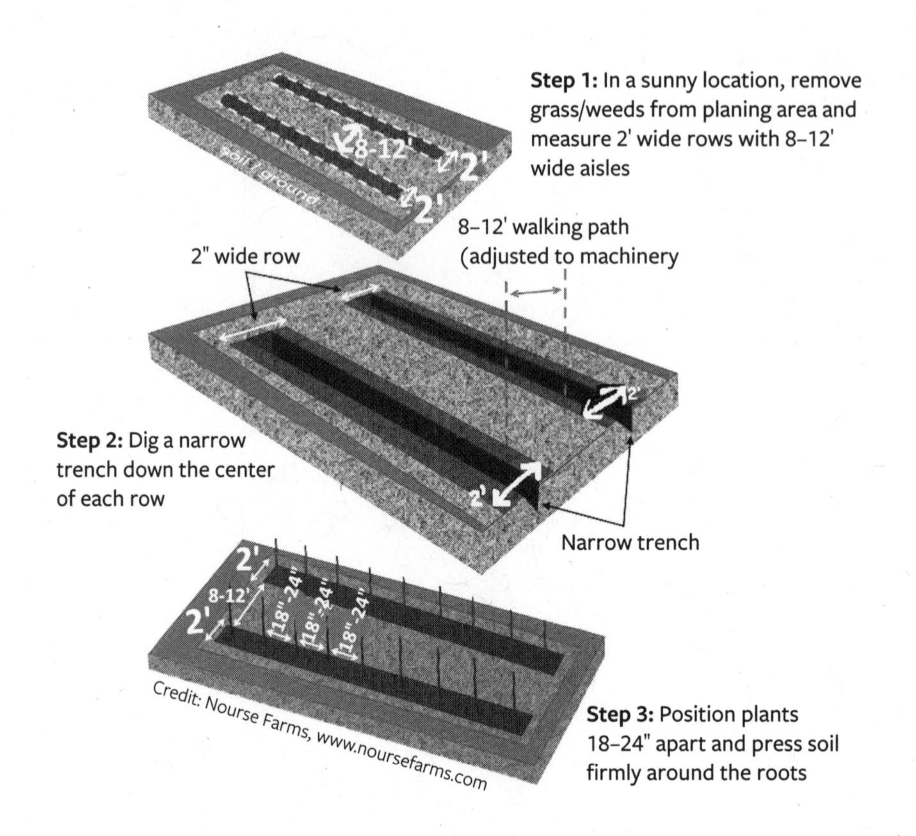

Foliar applications of kelp extract can be done any time berries are not ripe. Do not fertilize in late summer or fall.

Irrigation

Red raspberries are sensitive to drought but once established do not require irrigation except in dry climates or in drought conditions. Irrigation during initial fruit setting can increase yields, but too much near ripening will water down the flavor. Too much moisture, especially on heavier soils, creates a big risk for Phytophthora Root Rot (PRR), which will kill the planting. Utilize one or possibly two lines of drip tape per bed. Drip tape needs to be removed before mowing raspberries down to the ground in winter. Exact watering needs per season will depend on your region and soil type. Overwatering will kill them!

Black raspberries are drought hardy and once established do not require irrigation except in dry climates or in drought conditions to improve fruit quality. Irrigation during initial fruit setting can increase yields but too much will water down the flavor. Utilize orchard tubing with 2 emitters per plant or drip tape. Exact watering needs per season will depend on your region and soil type.

Pruning

June Bearers

For June bearers, the floricanes bear fruit after one year of growth, in June–July. After they have finished their summer crop, these canes are lopped down at the base and removed. Don't leave any stub. Leave the remaining primocanes to mature and bear a crop next year.

Fall bearers/Everbearing Raspberries

The berry crop borne on primocanes will finish by October (most regions) or early November. When the canes are completely dormant in winter or very early spring, mow them down as low as possible. Do this during full dormancy well before any new canes are emerging or green

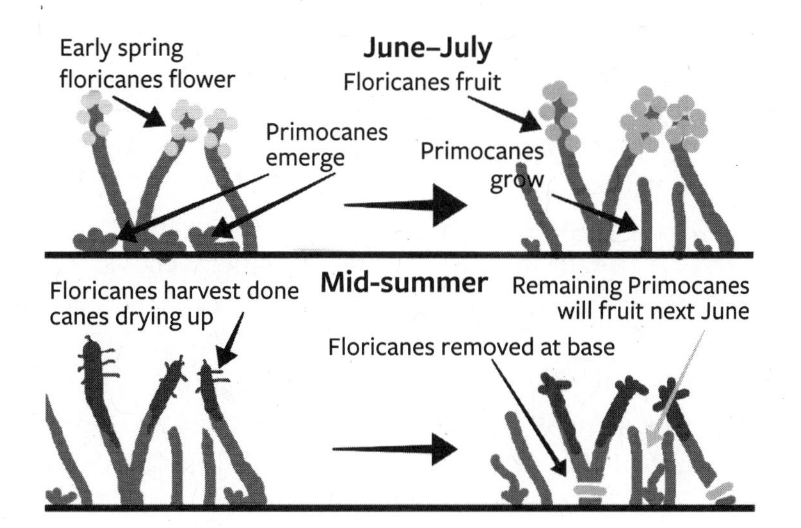

Pruning June bearing raspberries.

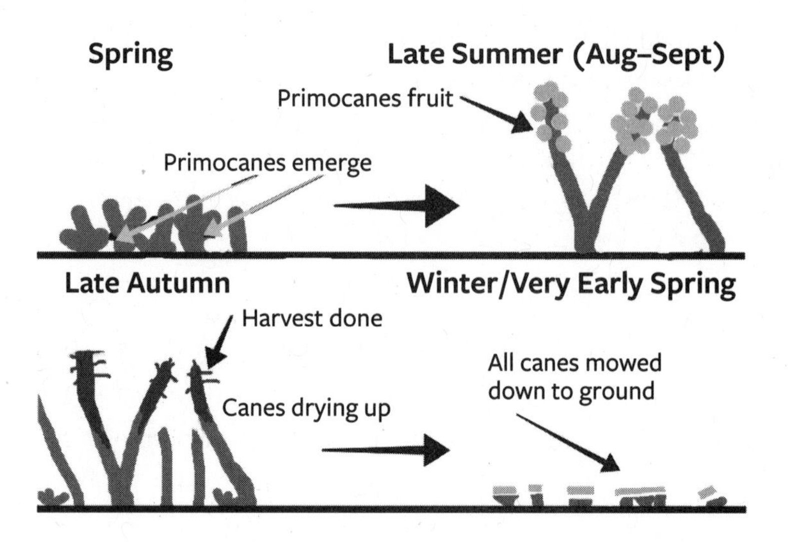

Pruning fall bearing raspberries.

growth is commencing. The primocanes that soon emerge in spring will bear the fall crop of berries, usually beginning in July or August.

Thinning

The canes of both varieties need to be thinned out or they will overcrowd each other and cause problems.

Thinning June Bearers

- After summer fruiting, the floricanes need to be removed completely (by snipping at the base) and then composted or burned. Don't leave them as mulch.
- The new primocanes need to be considerably thinned by removing about 50–75% of them. Leave the largest, healthiest ones in the row, removing all the spindly, crooked, and little ones, or ones coming up where you don't want them, such as walkways.
- Thus, thin them to about 6–8 canes per running foot of row. This is a bit laborious but is totally necessary. A 100' row would take about 2–4 hours to thin depending on your speed. Go quickly and don't over analyze. Good news: you only have to do this 1× per year.

- Wear gloves and long sleeves because the canes are prickly. Cut with sharp hand snips or perhaps use a *kama*, which might be faster.
- Remove any wilted, dead, diseased, damaged, or browning/drying up canes.

Afterwards, the raspberry patch will appear much more open, orderly, and attractive. Sunlight and air can now flow through unimpeded. This goes a long way to maintaining a heathy patch while keeping insects and diseases to a minimum. SWD cannot handle strong sunlight and likes humid, dark conditions, like an overgrown red raspberry patch. Keeping it very clean and pruned will help keep SWD to a minimum.

Thinning Fall Bearers

Fall bearers produce fruit on primocanes that sprout in spring and mature by late summer, at which point they flower and fruit. So, you need to thin out spindly, weak, crooked, or crowded canes in summer. Thin the same as June bearers. The main difference is after harvest in winter or very early spring you will mow the entire line of them down. Perhaps a powerful string trimmer might work too. You want to leave no stubs. It has been suggested *not* to mow down the entire patch of fall bearers until it is at least three years old so it can establish extensive roots, as mowing reduces nutrient absorption to some degree.[2]

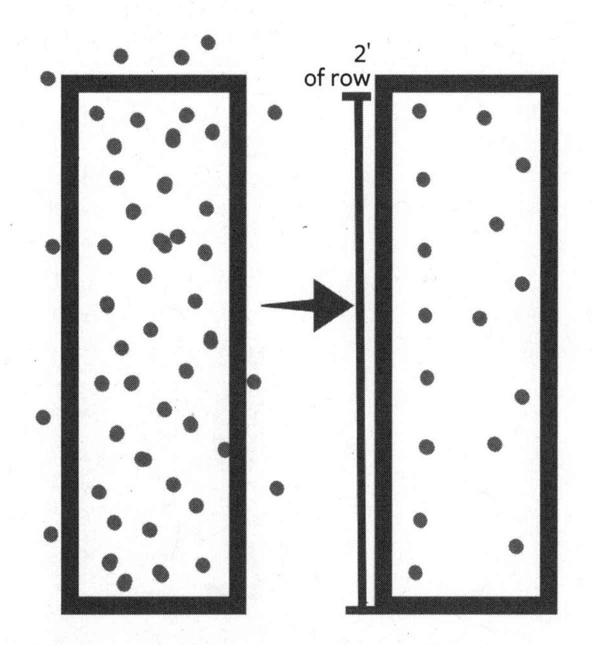

Thinning out shoots means removing all but about 6–8 per foot of row, and all that are weak or spindly or outside the beds.

Trellising

Red raspberries *have* to be trellised. If not, the canes will flop over and many berries will be lost.

The good news is that the canes are small and very light (unlike hefty blackberry canes) and so trellises can be light-duty and rapidly installed. We have found it very effective to put in 6' (1.8 m) tall metal (fence) T-posts about every 6' or so, driving them about 8–12" (20–30 cm) deep. Keep the "spine" of the posts all facing the same direction. Then take tomato twine or baling twine, or, even better, metal high tensile wire, and about a foot high up the posts, start the bottom line. Do a "Florida weave"[3] across every post and then back again. Tie off tight. Then do another Florida weave about 4' up, in the same way. Done. Twine only lasts a season, high tensile wire indefinitely. Make sure this gets done or your canes will be flopping over and snapping and yields will go way down. You may prefer to install a more permanent and sturdy trellis, such as a V trellis or cross-arm trellis.[4] Exact plans for these are easy to find from university publications. There are many designs and they are based on your training system more than anything.

Mulch

Red raspberries need mulch when establishing to conserve moisture and lower weed pressure. Thick straw mulch is ideal. Rotted tree leaves, rotted hay, grass clippings, or other high-carbon materials will work. Avoid wood

Credit: Nourse Farms, www.noursefarms.com

A modern high-density, super-productive raspberry planting showing close spacing and plants pruned short, around 4'. Note wooden posts used on the ends and high tensile metal wire.

chips unless very well-rotted. It's not recommended to mulch the plants after the first year or two of establishment.

Propagation

The fibrous, freshly cut roots can be used for propagation. Simply dig a shallow trench about 2–3" deep and 6 inches wide. In early spring, lay out moist, fresh roots into the trench, and immediately cover with about 2 inches of loose soil, and gently soak, then mulch. In 2–4 weeks shoots should appear. Or, you can start the roots in 72-count cell trays in well-draining potting soil and transplant the plugs 5–6 weeks later. Simply bury the roots half-deep down into the cell, cover with potting soil, and keep moist and warm.

Pests/Diseases/Challenges

Insects and diseases are mostly very manageable.

Rednecked borers (RNB) can be an issue and can kill off some canes. Remove any wilting canes at the base and destroy.

Stinkbugs can show up and will cause individual berry drupelets to bleach white. This lowers the appeal of the fruit but it can still be sold if it's not too apparent or widespread. Spraying for them is not realistic as the plants are in bloom while fruiting and any sprays can harm honeybees and pollinators.

Japanese beetles and slugs like to eat the foliage. Keep the patch well-mowed and clean. Sluggo® pellets keep slugs and snails at bay.

SWD can be a formidable issue on the fall crop, but in our experience *pruning for lots of light penetration and good airflow, upright training, and keeping the berries judiciously picked* prevents SWD from getting out of hand. Keep the patch free of rotting berries. SWD cannot handle strong sunlight, so keeping the canes carefully maintained, *thinned*, and trained is important.

Pyrethrums can help with many insect issues but *not* when fruit is ripening or the plants are blooming, and their bloom season lasts for most of the season. Honeybees and bumblebees really favor the nectar-rich blooms, and would be killed by using pyrethrum or spinosad (for SWD).

As far as diseases, starting out with healthy virus-free planting stock goes a long way. The main disease of concern is Phytophthora Root Rot (PRR), (*Phytophthora rubi*).[5] This disease causes wilting and dieback and can rapidly decimate an entire susceptible planting. Choose cultivars that are PRR resistant. Make sure red raspberries are planted *only on raised ground/raised beds and on well-draining sites.* Any amount of stagnant standing water on site after storms, even just for 4–6 hours, can initiate a PRR infection that will kill the planting.

When making beds we always mound the soil up about 5–6" (13–20 cm) similar to mounding rows for sweet potatoes, etc. This increases water drainage. If your site is on the wet side, be very proactive to create raised planting beds and plant PRR resistant cultivars, and/or install tile drainage pipes. Anthracnose can also be an issue but resistant cultivars, good airflow and training helps a lot.

Marketing

Red raspberries are still a rarity on the local food scene in most places, so it's a profitable wide-open niche. Who's going to take this one on? They may not garner as much enthusiasm *at first* as blueberries or strawberries, but once people taste high quality local raspberries, they will be hooked. So, give out free samples (following all local laws and regulations). I had a market customer who came and bought $20 worth nearly every single week. Distribute recipes for raspberry desserts, smoothies, jams, chutneys, proper freezing instructions, etc. Market them in ½ pint and pint clamshells *only* as the berries crush easily. In 2015 we were direct-selling them for $5 a ½ pint. Five dollars seems to be a friendly, easy price people like at markets. It sounds negligible but adds up fast actually. At $5 per half pint, that would be $12.50 per pound, which is a lot more than is usually estimated per pound for this crop. Typically, $1.50–3.50 per pound (450 g) is estimated, but with direct selling you can crush this conventional forecast.

Value-added goods include jams, jellies, fruit leather, wine (excellent), ice cream, chutneys, vinegars, frozen berries, and more. Restaurants might be interested. Selling to breweries could be challenging because they generally want very large volumes. Health food stores and co-ops would be interested if the berries were packaged professionally, clean, neat and hygienic, preferably with a sticker stating your farm name on each clamshell, etc. We sold to a local health food store about $50–100 a week in raspberries in addition to our market sales of about $100–200 per week off a tiny 100' (30 m) row, for about 10 weeks a season. That added approximately $2000 if not more to our seasonal market income, again from a 100' row of berries. It was very successful. It was also a new (1 year old) planting, on relatively poor soil giving mediocre yields, at a quiet farmers market and the demand exceeded our supply. We could have easily sold twice as many berries at our market, and 5–10× as many at a busy city market. For that we would have needed many more row feet of raspberries.

Also note, the *leaves* are a potent medicinal herb for rapidly alleviating menstrual cramps, and make a nice floral herbal tea. It could be worth looking into drying and selling the leaves as a product. Certainly, when you thin out all those shoots, you'll be left with a pile of them that could become either cash or trash. Maybe a local herbalist would be interested in that by-product. Plants are also in high demand.

There are hybrids of reds and blacks, known as purple raspberries. These include 'Brandywine' and 'Royalty'. Productive and tasty, but the soft fruit is likely best used for home plantings.

In conclusion, red raspberries are one of the easiest, most productive, and quickest to profit small fruits for the market grower. They need minimal space, even 50 or 100' of row can be very productive on good soil and bring in about $10–40 per foot of row if not more. Perhaps the easiest small fruit to grow and market besides tomatoes. Excellent for backyard growers as well. Very underrated.

Cultivars

Reds (June bearing/floricane)

AAC Eden: From Nova Scotia, this cold hardy, spineless cultivar produces mid-season harvests of very good flavored, firm berries.
Boyne: Productive and winter hardy, short plants. Produces small to medium sized berries of good quality. From Geneva, NY.

Cascade Premier: Released in 2022 by Washington State University. Has large-size fruit with excellent flavor, appearance, production, and strong disease resistance to PRR. Bred for the PNW. Patented.
Dorman: A wineberry (*Rubus phoenicolasius*) cultivar masquerading as a red raspberry. Produces mediocre berries where it's too hot to grow real raspberries: the Deep South, Florida, and Texas. Home use only. May have potential as a processed food.
Encore: Late midseason ripening extends the season. Very cold hardy.
Nova: Mid-season ripening. High quality berry. Easy to grow because it adapts to more zones than any other variety.
Prelude: Large berries, early summer bearer and also possible to get some fall berries. Heaviest production is the early summer crop. Very good flavor.
Tulameen: PNW cultivar with excellent, flavorful, large fruit and production. Prone to PRR.

Reds (Fall bearing (everbearing)/primocane)

Heritage: One of the most popular backyard cultivars. Productive with tasty berries. Not the best for marketing purposes, better cultivars are out there.
Caroline: Combines disease resistance (PRR resistance) with earlier ripening in late summer, about 1–2 weeks before 'Heritage'. Produces berries with excellent, tangy-sweet flavor and good production. In our trials the berries are very tasty but somewhat small and soft. Best for home use.
Anne and **Double Gold:** Beautiful golden yellow fruits with high quality.
Mapema®: Early producing primocane berry with heavy production, very good quality and excellent flavor. Zone 5–9 (moderate summers only). Patented.
Kweli®: Heavy yielding and good shelf life due to drier berries. Handles high temperatures, being grown even in hot dry tropical locations. Patented.

Blacks

Jewel: Very high quality and large berry. Very productive. Mid-season.
Bristol: High quality, delicious berries, strong early-season production. Very popular.
Mac Black: Late-season berry with high quality and large berries.
Niwot: A newer, primocane raspberry that can produce berries in both early summer and again in late summer/early autumn. This one is impressive. After a good spring crop in June, heavy primocane yields come on in August-September on the *tips of the primocanes.* That means you cannot tip-prune this cultivar or you risk losing the primocane crop, which basically doubles the yields. This vigorous cultivar *requires trellising* of some kind to handle the spindly 6–7' (1.8–2.1 m) tall fruit bearing primocanes. After harvest you can prune it shorter to 4–6' (1.2–1.8 m) tall.

Summary

Red raspberry is *the top choice* for both the urban and rural market farmer where summer temps are not too extreme. Best in mild to moderately hot summer locations only. Easy to grow, minimal upkeep and minor infrastructure needs. Resilient cultivars are resistant to pests and diseases. Highly productive of high-value fruit. Space requirement is small. Production starts in only 6–12 months from planting. Black raspberry is tasty and a good crop, but not nearly as productive or in demand. Primocane black raspberries are much more productive. *Profit potential*: For reds, around $5–15 per cane, 6–8 canes per foot.

Urban Market Farming Rating—5/5: In appropriate climates red raspberries can be highly productive and easy to grow. Highly efficient on a dollar yield per square foot basis, especially compared to other fruits. Small space requirements, strong production, very high demand and price, as well as quick production time (6–12 months) make this the *best overall small fruit for the urban grower to get into.* Choose a good quality site as outlined earlier and appropriate cultivars of both June and fall bearing raspberries. Annual maintenance is important. Black raspberries are not generally space efficient or profitable for urban settings.

Rural Market Farming Rating—5/5: Easy to grow and highly productive. Make sure quality is high and local demand is there. Weed and grass control plan must be in place. Black raspberries can be profitable in areas where they are popular if planted in mass. Both would be good for You-Pick.

Home Recommendations: 20–50' of red raspberries will produce an abundance of tasty healthy fruit for everyone. Make sure to plant on slightly raised beds in a sunny to slightly shaded location that does not hold water. 'Heritage' is the most popular home cultivar, but many others work well. Black raspberries are great in sunny or shadier locations, OK near walnut trees or even in herb gardens. Make sure fertilizing, trellising, heavy mulching with straw/grass, and thinning/pruning of canes is done diligently every season. The leaves of red raspberry are medicinal and once dried are good in tea.

Identifying Quality Stock: Red raspberries should have lots of fibrous roots. You may see tiny whitish buds on the roots, these are the future canes. The stems should be pencil thick, or smaller is fine as long as there are lots of roots. The roots should not be rigid and fixed in place, this identifies an older plant that will not perform as well. Black raspberries have much less fibrous roots but should be similar. Plugs and root cuttings also work very well for both.

13

Strawberries

Nothing welcomes in spring vibes like cheerful, fresh sun-ripened strawberries! The market grower will have the advantage over the grocery store with the intense flavor and aroma of local strawberries. The best strawberries are not the huge deformed red blobs you see in the grocery store, which are rapidly decreasing in flavor. The best strawberries are medium to small, dark red, and intensely flavorful, tangy, and firm-soft. They get a great price at the market and people enthusiastically buy these vibrant (often) first-fruits after the close of a long, fruitless winter.

Cultivation

Strawberries are easy to grow and great for beginner growers. In fact, they were the first fruit I ever grew back in the very early 2000s. You always hear they need an acidic soil, but unless your soil pH is very high (alkaline) or limey, you probably will do fine without trying to purposely turn the soil acidic. Altering soil pH is risky business. Just provide a richly amended, loose bed and mulch if able. Dried pine needles, straw, or oak leaves are ideal mulch. They thrive with high organic matter, compost, and fairly fertile conditions. They need excellent soil drainage and so avoid poorly-drained areas and frost pockets. Slightly raised beds work great.

If your soil is very heavy clay or infested with rhizomatous weeds you may need to cover crop a season before planting for best results. Some cultivars grow better on clay soil than others, so inquire with your local

Organic strawberries being grown on slopes in Germany. Note terraces made with block and landscape fabric for weed control. Black currants in center background.

ag extension office and nursery. Avoid sites where nightshades and raspberries grew recently, as the soil may harbor harmful verticillium pathogens. Plant in full sun or nearly full sun locations.

There are 2 main types of strawberries: June bearing and everbearing. For marketing purposes, focus on June bearing cultivars which produce a lot of berries all at one time, which is what you will need to bring quantity to market. You can stretch out the strawberry season if you want by planting early, mid, and late season berries. Later season strawberries will usually suffer more from insects and diseases than earlier ripening cultivars.

Growing Methods

Basically, there are 2 main methods to growing strawberries. First is the annual production (AP) system. Second is the matted-row (MR) system.

For both systems, make beds about 12–18" (30–46 cm) wide and leave about 18" between beds of berries.

AP: Plant bare-root everbearing starts in early spring (or autumn in mild winter areas) and harvest the berries that same spring. Plants are removed/tilled in after harvest and thus treated as annuals.

Advantages: Easy, very fast production. Less care and maintenance. Less opportunity for weed takeover or disease build-up.

Disadvantages: Smaller yields and slightly lower quality fruit.

MR: Bare-root June-bearing starts are planted in early spring (or autumn in mild winter areas) and you establish and maintain the plants that year, removing all blooms and weeds, allowing no fruit production. In early summer, arrange the abundant suckers that have formed to fill out the bed (usually 4 per "mother plant"), thinning out the rest, and keeping them all about 1' apart.

You harvest the berries the following spring/early summer (one year after planting). Plants are thinned to 1' (30 cm) spacing after harvest or can be mowed to 4–5" (10–13 cm) and thinned later in the season. Do not mow or damage the crowns in this process or you can ruin the patch—set your mower blades high enough! Plants are treated as perennials and left to bear fruit for 3–5 years.

Advantages: Larger yields of better berries. Plants yield for a number of seasons and not just once. Better return for labor overall. Carefully thinned plants are also a viable and profitable product.

Disadvantages: Much more labor is required. Weeding strawberry beds is challenging. Removing all the initial blooms is challenging on larger plantings. Plants must be thinned every season. This is labor intensive, but could also be a substantial source of additional income. Diseases and viruses can build up over time. Labor is easily tripled or quadrupled but so are yields, and could be worth it for you.

You'll have to decide what is important to you and how much time and labor you have to dedicate to strawberries. Many growers are going with the AP system. Backyard and micro growers might prefer the MR system.

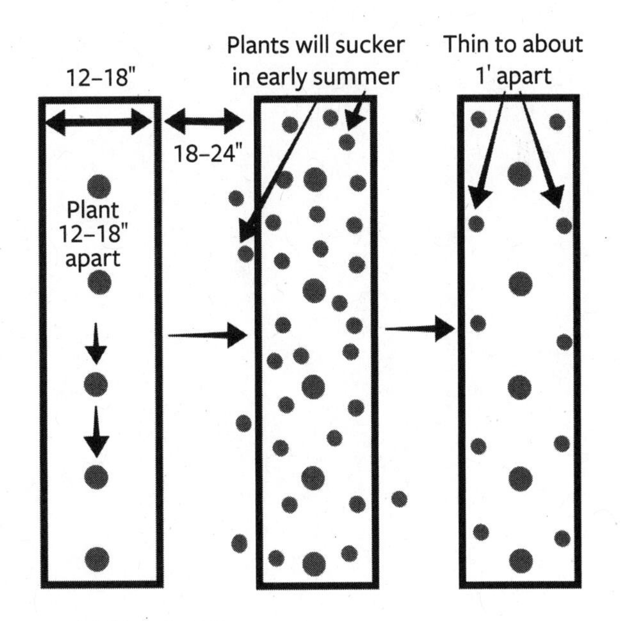

Establishing the MR system.

Thinning

Thinning out the plants is important in the MR system and can be done anytime between early summer the year of planting, to early spring, before flowering starts. This means removing many, or most of the plants and runners so plants are 1' apart. Strawberries need good airflow and sunlight penetration, both of which are crucial to keep the berries from molding, and which unthinned beds will be lacking. Even small beds can produce hundreds of excess plants, so thin judiciously. Those thinned plants work great to start new patches, fill in gaps in the bed, and also are a valuable commodity in themselves. If you have time, thin the suckers in late winter or early spring, wash off the roots, bundle and bag them in bunches of 10 and sell them at the market for $5–10 each or more. Many bags would sell and it'd be better than tossing them all out. Also they can be potted and sold.

MR bed on our farm in June a few months after planting shows the bed already filled with lush suckers. This will fruit next May.

It should be noted that many commercial growers are currently experimenting with growing strawberries in containers, even vertical ones. This may be something to look into.

Planting

Plant each strawberry plant 1' apart in the bed. Plant 2 off-set rows of strawberries across a 30" (76 cm) or wider bed, 1 foot apart. Or, if utilizing 18" wide beds, plant one single row down the middle planted 1' (30 cm) apart. This will seem excessive spacing at first, but they fill out via suckers rapidly. Don't plant them closer than 1' apart.

Plant in the correct season. If you live in *zone 7b or colder*, plant in spring only. Deep freezes in winter will often kill autumn plantings, thus the need for spring planting. In milder areas, autumn planting works well. Plant in early spring, whenever the weather is settling out, around late March–mid May is best, earlier in Southern areas. Frost does not harm the new plants. Don't plant after June 1st in most areas.

Plant the correct depth and spread the roots. You will notice strawberries have a tiny stem; a *woody mid-point in between the roots and the leaf stems.* This is called the *crown*, and it cannot be buried or the plant will fail. Also, if strawberries are planted *too shallowly* and roots are exposed the plant will fail. So, you must plant at just the right depth to have roots buried but the crown above the soil line. After planting and watering in, sometimes the soil is washed away from the roots, so make sure you plant deeply enough and are gentle when watering new plants in. Then mulch. Re-cover any exposed roots with soil and mulch.

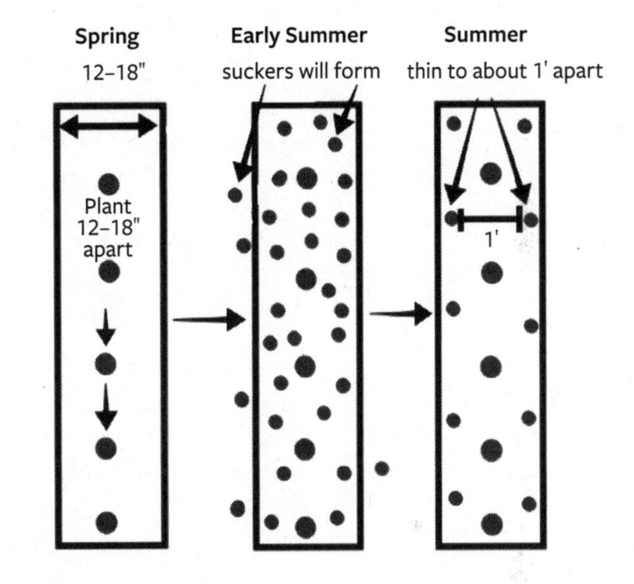

Mulch. Straw, grass, or rotted wood chips work well. Strawberries respond to compost added to the soil or on top before the mulch is applied. These days many growers grow them on raised beds on black plastic, with a small hole cut or burned on each spot where a plant will be planted. This reduces weeding but is an additional expense. Look into biodegradable options. Some use landscape fabric.

Remove Flowers. Traditionally growers remove the flowers the first season and this allows the plants to establish better and produce a larger 2nd year crop. This is not done in the AP system.

Irrigation

Strawberries are quite sensitive to drought once established but do not require regular irrigation except in dry climates or in drought conditions. Irrigation during initial fruit setting can increase yields but too much will water down the flavor (very bad) and can rot or split the fruit. Stop irrigation as soon as fruit starts coloring up. Overhead irrigation during morning hours is fine as long as the fruit is not ripening. Drip tape can also be used, 2 lines per bed. Exact watering needs per season will depend on your region and soil type.

Harvesting

Berries start ripening around mid-May to June in most areas, earlier down South. Once berries start showing pink or red coloration, harvesting will need to begin within a few days. Harvest and handle the berries carefully, possibly even with cotton gloves, to avoid puncturing or bruising them. Keep the green calyx on the berries. Stack no more than 1 or 2 berries high. You may want to harvest directly into clamshells. Wet berries mold rapidly, so harvest after dew has dried but before the daytime temperature gets hot. As you harvest, remove damaged/moldy berries by picking and placing them into a "moldy berry" bucket and dumping or composting well away from the patch. Immediately place the harvested berries in cold storage until market day. A patch needs picking at least 2–3× per week in the peak season.

Marketing

Strawberries sell themselves. Pack into pints and quarts. Value added potential is high: frozen berries, jams, syrups, restaurant sales, shortcake, etc. Direct selling these luscious fruits at a market should be easy unless competition is prolific. Make sure berries are fully ripe, clean, free of bruises, insects, and mold and the tops should be bright green. Offering free samples of your superior organic quality could pay off.

These are a popular crop for You-Pick.

Pests/Diseases/Challenges

Thankfully, strawberries ripen so early in the season that they do not have too many insect issues. Slugs/snails, pill bugs, and stinkbugs can all take interest in the fruit. Keeping the plants thinned out and free of weeds and decaying fruit reduces pest pressure.

Mollusks: Iron phosphate (Sluggo®, etc.) pellets broadcast into the patch will keep mollusk levels very low and may be needed in wet seasons and in dewy, moist fields—like ours! *Reapply the pellets as necessary, usually 1x per week if pest pressure is high. After blooming is fully completed, you may apply pyrethrum for pill bug and stinkbug control. Never spray the blossoming plants with insecticides that could harm honeybees and pollinators.*

Stinkbugs: Feeding by stinkbugs can cause a deformity of the fruit known as "cat facing," where the folded-in berry resembles a cartoonish cat face. This lowers the marketability of the fruit, but a few here and there should not present an issue for the market grower.

Deer: You may need to net the planting or fence if deer pressure is very high.

Frosts: Late spring frosts can damage the sensitive blossoms. These are one small fruit that is not frost tolerant. Damaged blossoms may still have white petals but upon inspection will show a black center point where the ovaries have been destroyed. Small plantings can be covered the afternoon before a frosty night with heavy row covers (you may need more than one if it is getting colder than the cover is rated to protect above freezing). Larger plantings are protected with overhead mist irrigation going all night.

SWD: These are pests of late season or everbearing strawberries only, so it's recommended to mostly focus on June bearing cultivars.

Diseases include gray mold and common spot. **Common spot** (*Mycosphaerella fragariae*) on disease-prone cultivars can be a problem, but most disease-resistant cultivars handle it well. Susceptible cultivars lose their leaves and fruit. **Gray mold** (*Botrytis cinerea*) can reduce yields in very wet spring conditions. It appears as a fuzzy gray mold on the berries and can spread rapidly via spore release. If you see it developing, try to remove any moldy berries right away. Harvest them in a separate bucket and bury or put in the middle of a compost pile. Gray mold pressure varies by year; in 2019 we lost maybe 10–20% of our strawberries, in 2020 and 2021 it was hardly present. Maintaining open, proper spacing of plants (1' apart), keeping fruit regularly picked and keeping the patch clean of rotting or moldy fruit helps keep it under control. Organic fungicides for gray mold can be researched and are available. **Red Stele** (*Phytophthora fragariae var. fragariae*) and **Verticillium wilt** can be serious diseases also, so utilize *resistant cultivars*, and plant on sites with well-draining soil only, where raspberries and nightshades were not grown previously.

Rotate planting sites; strawberries are viable on a site for *no more than about 5–8 years*, at which time you should mow and then smother out the plants with a tarp or till under.

Cultivars not suited for organic production include: 'Polka', 'Korona', 'Dania', 'Tenira', 'Rafzusen', 'Senga Sengana', and most major commercial strains. See below for recommended cultivars for organic production.

Cultivars

Allstar: Large, high-quality berries. Adaptable and disease resistant.
Earliglow: One of the earliest ripening cultivars. Highly disease-resistant heirloom. Excellent quality with fantastic flavor and tanginess. Medium-sized berries, some get pretty plump. Size decreases as the season progresses. Good for organic market growers. Ripens around May 15th in KY. Good yields.
Florence: Disease resistant. Good production and quality, large berries. Late-season.
Honeyoye: High quality, early-ripening berry. Considered an excellent choice for organic growers. Early mid-season.

Jewel: Big, beautiful, firm berries with very good flavor, early ripening about 1 week after 'Earliglow'. Consistent large size and high quality; very marketable modern look. Resistant to disease so good for organic production.
Surecrop: Disease-resistant and reliable heirloom cultivar. Very good flavor, medium-sized berries. *Blooms late* and is mid-season ripening.
Galletta: Early season production of high-quality fruit. Good for organic production. Patented and released by NC State University.
Annapolis: Early season and *very winter hardy to zone 3*. High quality berries.
Cavendish: Good for Northern areas. High yields over a long season. Disease-resistant. Not for hot areas. *Zones 3–7*. Early mid-season.
Flavorfest: Recommended for organic production. Large, flavorful berries. Does best on clay soil. Good reports from local growers on this one. Mid-season.
Yambu: Vigorous and productive, long season, good quality. Disease resistant. Early mid-season. Patented.
Chandler: Popular early-season cultivar. Productive of big berries. Flavor good-mediocre. Said to increase in quality as weather warms up.

Summary

Extremely popular fruit and very high value crop. Perishable and susceptible to excess rain damage and late frosts, so a bit risky. Make sure to plant on raised beds and have frost protection blankets on hand in case of late frosts. Harvesting and maintenance are fairly labor-intensive. Overall easy to grow. Beds or rows last about 4–8 years before they need replacement in the MR system.

Urban Market Farming Rating—4/5: Growing them in an urban setting may not be efficient on a dollar yield per square foot basis compared to more profitable crops such as tomatoes, greens, etc. However, the high demand and good market value of the fruit, small size of plants, quick production time (2 months to 1 year), and ease of growing may make it worthwhile. Vertical growing in containers might be worth looking into.

Rural Market Farming Rating—4/5: Fairly easy to grow, and without space restrictions this could be an excellent niche crop to get into. Make sure quality is high and site selection is good. Harvesting and also weed and grass control can be challenging. Popular at You-Picks.

Home Recommendations: Start with healthy, virus free plants and plant early in spring into fertile soil in prepared beds. Make sure plants are thinned in winter/early spring to 1' apart, before flowering commences. Thinned out plants can be utilized for a few years to start new patches. Replace a patch after 4–5 years of fruit production. Keep weeded. Developing berry clusters can be "hung" over the leaves to keep them off the soil, drier, and cleaner. Keep thick row cover on hand to cover during flowering if frosts are coming. A row of 20–50' will abundantly supply most homes. Excess berries freeze well and make great jam and icecream.

Identifying Quality Stock: Lots of fibrous roots and crowns that are intact and not mushy, broken, or soft. There should be small emerging leaves at the top of the crown. Bare-root is always best. Source virus-free plants.

Michael Hicks, owner of Living Roots farm in Indiana, had these comments on his small-scale organic strawberry production:

> The biggest challenges are weed control and a short harvest season. Most large market growers use black plastic, but we won't, so that causes more weed problems. When we were growing the most, we grew 'Earliglow', 'All Star', and 'AC Valley Sunset' (early, mid-season, late season, respectively). I like 'Earliglow' and 'All-Star'. The 'AC Valley Sunset' would get more bug issues coming later (and also more rot as we got into the heat of June).
>
> Strawberries sell well for a decent price at a good farmers market ($6/quart), but most of ours were for our You-Pick. Most You-Pickers would come out on the weekends. If you got a rainy weekend, many berries would be lost to rot. So, you had that risk

> too. They also come on when we are the busiest getting summer crops in the ground, so that timing isn't great. Another issue we've had is having to cover and uncover several times throughout the winter—since the weather has been erratic. For the amount of work (weeding mostly), the profit is less than any other produce item on our farm. We followed my grandfather's (Stanley Hicks) method. After harvest season is over, mow the strawberries down to about 4–5 inches (don't cut into the crowns!). Then thin the patch to one plant every foot. This helps with disease and rejuvenates the patch. Despite the challenges, we still grow strawberries, just less than before, and mostly for our CSA.

Among other fruits, Michael also produces 'Chicago Hardy' figs inside a high tunnel, and considers these a high-value crop worth growing.

14

Juneberries

Imagine juicy, flavorful blueberries with tiny almond-flavored seeds on a large leafy shrub and you have an image close to Juneberries. Truly, the ornamental, native shrubs are beautiful and the very nutrient-dense fruit can be amazingly delicious, at best a complex, rich sweet flavor reminiscent of dark cherries, almonds, concord grape, and blueberries combined. Mediocre varieties are small fruited, watery and bland, as well as susceptible to common diseases, so cultivar selection is important, as well as *species selection*, as we will explore. They are also called *serviceberries*, *shadbush*, and *saskatoons*, and this varies with region and species. They are more nutritious than most blueberries.

Recently these grossly under-utilized pome (apple family) fruits are gaining a lot more attention and research, especially in New England, and in the upper Midwest and bordering regions of southern Canada, where growing other mainstream berries commercially can be challenging or impractical due to extremely severe winters and other factors. Canada currently has over 900 farms growing juneberries on over 3200 acres.[1] Elsewhere they are being seriously evaluated and trialed as a new berry crop, as they absolutely should be. A professional, commercial production manual can be ordered from the government of Alberta, Canada.[2]

Cultivation

Juneberries are relatively easy to grow and are adaptable to various soil conditions (unlike the similar, yet unrelated, blueberry). Loam and clay soils that are very well-draining are ideal. Sandy soil would be fine too with lots of mulch and compost added on top. Good drainage is required. The flowers do show some level of frost tolerance but avoid frost pockets. Although in the wild they tolerate shade and partial shade, plant in full sun locations only for fruit production.

Beds are not conducive or needed for this large shrub. Planted rows are generally 12' (3.7 m) apart, and 4' (1.2 m) between plants. (Rows could be closer if not mowing with a tractor.) Plant bare-root, dormant plants early in the spring season (potted is also OK, just more difficult to handle and more expensive). Juneberries are drought resistant but need irrigation when establishing, and may need to be irrigated a few times per season depending on local rainfall. Juneberries are not adapted to desert or subtropical/very low chill regions.

Establish healthy bushes by eliminating all grass and weed competition for the first 3–4 years, similar to the former fire-cleared prairies where this species originated.[3] This is best done by removing any sod at planting and keeping plants deeply organically mulched and heavily weeded, or utilize landscape fabric. If you use plastic you must make sure the holes are large enough to facilitate suckering, as juneberries sucker similar to blueberries yet get even larger. Shallow cultivation can be used to help with weed control, but deep cultivation can damage the roots.[4] Annual spring and early summer applications of a balanced organic fertilizer will ensure strong growth and production. You can keep the bushes/small trees pruned to make netting and harvesting easier.

Juneberries bloom in early spring, with fruit forming six to eight weeks later. Plants begin to bear fruit two to four years after transplanting. Significant yields can be expected after six to eight years, with maximum yields after 12 to 15 years. Mature plants may yield 3 to 15 pounds of fruit per shrub, depending on species and size. Well-maintained plantings can be productive for 30 to 50 years.[5]

Clusters of fruit ripen fairly uniformly, making it possible to harvest the crop within a short window of time. Due to the slightly lower water content, juneberries that have been refrigerated immediately after harvest keep for 7–14 days if covered.[6]

Juneberries come in two widely grown species. The species to grow for berry production in northern regions, as well as along the PNW is *Amelanchier alnifolia* (saskatoons). The native Eastern US tree-form serviceberry is *Amelanchier canadensis. Amelanchier canadensis* and *Amelanchier x Grandiflora* are likely the best species for plantings in *warmer and southern regions due to having better disease resistance needed in these regions.* Make sure any plants you source are the *correct species for your region* and should be cultivars that produce tasty marketable fruit. Seedlings of good parentage can also be utilized.

Consult with regional growers on species selection. Planting the incorrect or non-optimal species for your area could result in failure or major disease issues and crop loss. I've experienced this. Cultivars can be challenging to source at this time, but make sure you stick to the recommended species for your region! Canadian nurseries might be the best source for superior cultivars at this time. An in-depth online grow guide can be found here.[7]

Seedling Juneberry with excellent quality fruit, found in a parking lot landscape in Louisville, KY. We came back in winter and took cuttings and suckers!

Best species by region

Northern California, PNW, Canada, Prairie, and upper Midwest: *A. Alnifolia (Saskatoons)*

Mid-Atlantic and Southern USA: *A. canadensis* and *Amelanchier x Grandiflora* (for best fruit production); *A. Laevis* and *A. Arborea* also grow well but are more for wildlife.

New England: *A. canadensis, A. Arborea,* and possibly *A. Alniflia*

Amelanchier alnifolia matures to about 18' (5.4 m) tall, but growers should keep them pruned to about 6–8' (1.8–2.4 m) tall for easier *bird protection* and harvesting. *A. Canadensis, Amelanchier x Grandiflora*, (and to a lesser extent for fruit production *A. Laevis, A. Arborea*) are perfect for areas such as the upper South, Mid-Atlantic, and the Ozarks where intense cedar apple rust (CAR) pressure and strong summer heat *diminish the usefulness of CAR susceptible A. alnifolia as a crop. A. Canadensis, Amelanchier x Grandiflora, A. Laevis, and A. Arborea* are not susceptible to CAR, and some appear nearly immune.

Seedlings (of a regionally-appropriate species) with good genetics can often reliably produce high quality, marketable fruit, so named cultivars may not always be necessary. Seeds require cold stratification (32–39°F, −5 to −3.5°C) for 90–120 days before planting.

Irrigation

Juneberries are resistant to drought once established and do not require irrigation except in dry climates or in drought conditions. Irrigation is beneficial during establishment years. Irrigation during initial fruit setting can increase yields but too much will water down the flavor. Overhead irrigation is not recommended. Stop irrigation as soon as fruit starts coloring up. Utilize orchard tubing with 2–4 emitters per plant. Exact watering needs per season will depend on your region and soil type.

Pests/Diseases/Challenges

Juneberries are fairly resistant to insects and diseases but suffer a few of the same maladies as apples including *cedar apple rust* (CAR): *Gymnosporangium juniperi-virginianae* and other *Gymnosporangium* species that send forth airborne spores from juniper ('cedar') tree-borne galls. CAR is very problematic and, in many seasons, will ruin some portion of the fruit—sometimes 50–75% or more in a bad season on susceptible cultivars/species. CAR covers the fruit in tiny yellow spikes (a form of sporulation structure) and also similarly harms the foliage. Removing all nearby juniper/red cedar trees, if possible, within a ½ mile helps, and many urban plots may have less CAR pressure. Those same cedars would make great posts for fencing and supports for the required bird nets. *If CAR is a serious problem in your area on apples or juneberries, then only plant A. canadensis and Amelanchier x Grandiflora, A. Laevis, A. Arborea.* In the Mid-Atlantic and South this is necessary as juniper trees are often abundant.

Credit: Wikipedia

Cedar apple rust is a big problem so choose species and cultivars carefully.

Fire blight (*Erwinia amylovora*) can also be an issue, again similar to related apples and pears. Treatment is essentially the same as for other fruits and focuses on pruning out diseased twigs and branches back to healthy growth. During the dormant season (late autumn to very early spring) prune off and destroy all diseased branches, cutting 10–12" (25–30 cm) below the infected area. All pruning tools need to be sprayed with isopropyl alcohol in between cuts so as not to spread infection. During the active growing season prune off and burn all infected twigs and branches in the same manner, cutting well below the infected areas. Any bushes that are severely infected with large fireblight cankers in the trunk should be immediately removed and burned so as not to spread the disease. Fertilizing apples, pears, and juneberries with excessive nitrogen, especially synthetic fertilizer, makes them much more prone to this disease.

Mummyberry can also be an issue, treat the same as with blueberries.

Entomosporium leaf spot (*Entomosporium sp.*)[8] *Entomosporium* usually appear as small, dark circular spots on the leaf, usually with a halo outline. There are copper and also sulphur fungicides that are registered for use on rusts that might be an option. Avoid overhead irrigation and make sure airflow is good.

Insects are minimal but include **apple curculio** (*Tachypterellus quadrigibbus*). Pheromone traps and sticky traps can help. Avoiding large monoculture plots should keep most insects from becoming major issues.

Birds pose a significant challenge, often stripping bushes clean of fruit, so bird netting is absolutely necessary and needs to be up directly after bloom is completed.

One of the best aspects of this cold-hardy crop are that the buds, blooms, and young fruits are *very resistant to late frost damage*. Frosts do not destroy the flowers at the time of bloom on most species.

Marketing

In northern areas these berries are gradually catching on and should not be too challenging to market. Pack "prairie blueberries" in ½ pint and pint clamshells. Handle carefully as the berries are quite fragile and soft, being more perishable than blueberries. In other areas public education campaigns and taste testings will improve market viability. The berries are delicious and so free samples should generate sales. Juneberries go by a few different names ("sarvis berry" in rural KY, saskatoons in Canada, etc.) so be sensitive to culture. Value-added goods have potential: jams, pies, wines, fruit leather, etc. The seeds and plants (from suckers or cuttings) are highly marketable.

Juneberries on our farm safely ripening under bird netting. Note that the berries are not pressed against the netting.

Cultivars[9]

Regent (*A. Alnifolia*): Popular and widely available saskatoon, but mediocre fruit quality and lower yields than others. Not very marketable.

Martin (A. *Alnifolia*): A preferred commercial saskatoon cultivar. Large, sweet berries and heavy production.

JB-30 (A. *Alnifolia*): Newer commercial saskatoon cultivar with large, excellent berries and less suckering.

Honeywood (A. *Alnifolia*): Compact grower, produces large harvests of sweet, quality fruit.

Smokey (A. *Alnifolia*): Large, fleshy fruit, sweet, mild flavor, very productive, medium-sized clusters.

Pembina (A. *Alnifolia*): Large fleshy fruit, delicious sweet flavor. Grows in long clusters.

Northline (A. *Alnifolia*): Similar to 'Pembina' in quality and most aspects. Ripens unevenly.

Autumn Brilliance (A *x grandiflora*): Purple fruit with good flavor. Fruits are ⅜" (0.95 cm) in diameter; plants reach a height of 20 to 25 feet (6–7.5 m) at maturity. Fairly resistant to diseases including leaf spot and fire blight. Grows and produces well in the Mid-Atlantic.

Ballerina (A *x grandiflora*): Purple berries when ripe and ⅜" to ½" in diameter. University of Kentucky growers rate this fruit as "superb, large, sweet and juicy." This plant reaches a height of 15 to 20 feet. It is hardy as far north as zone 4. Good resistance to leaf spot and fire blight. Challenging to source. Maybe contact the University of Kentucky.

Prince Charles (A *x grandiflora*): Fruit are ¾" (1.9 cm) in diameter and rated high in taste evaluations in Georgia. Bushes are upright and reach an estimated height of 25 feet.

Princess Diana (A *x grandiflora*): Juicy, blue-black fruit are juicy and sweet. Fruits reach ⅜" in diameter; plants can reach a height of 25 feet; good leaf spot resistance.

Summary

Productive, super delicious, and very healthy native fruit. An ideal crop for northern and prairie regions that lack other options, are not conducive for blueberries, and/or that have an extreme winter climate. Also, a more adaptable and viable alternative to blueberries. Juneberries are easy to maintain, mostly disease and insect resistant and fairly productive. The fruit is still little-known in many areas but is increasing in popularity and is bound to catch on and could be a fairly easy sell once people taste the fruit. The size of the bushes and relatively light production makes them more suited to rural farms. Birds pose a significant challenge and so netting is absolutely necessary. Best grown as a pruned bush and not left to grow into a large shrub or tree form. *Verify you are choosing the correct species for your region before purchasing or planting any stock.* Profit potential: Around $30–50 per bush.

Urban Market Farming Rating—1/5: Not very space efficient per pound of fruit produced. Demand is currently low but this will improve in time.

Rural Market Farming Rating—4/5: Easy to grow, without space restrictions this could be a good niche crop to look into, especially in northern and prairie regions. Choose the most disease-resistant, high-quality cultivars possible. Harvesting by hand is not practical on large plantings.

Home Recommendations: These fit well into landscaping and areas with good sun exposure. Keep pruned short so birds don't get all the berries and be prepared to net them. Make sure to get the correct cultivars/species for your location. They can often be found growing in commercial/industrial landscaping if you look carefully. Why let all that good fruit go to waste? Cuttings can be obtained from superior landscaping specimens also.

Identifying Quality Stock: Plants should be at least 1–2 ft tall and have lots of roots. Tiny plants take a long time to establish and have a poor survival rate. They all require careful irrigation and weed control while establishing.

15

Muscadine Grapes

Truly a Southern delicacy, I'm often surprised how many people have never heard of, or eaten, muscadine grapes (*Vitis rotundifolia*). Most people living south of Kentucky are quite familiar with muscadines. For the most part, Kentucky is just north of where muscadines feel comfortable enough to grow (zone 7). However, Mid-Atlantic areas closer to the ocean, yet further north, can usually grow muscadines.

Personally, I consider muscadines the best all-around organic market growing grape, and that's why I am focusing on them exclusively. Maybe this is just my Southern bias. Other species of grapes are, of course, good but are plagued with so many challenges for the organic market grower as to make them impractical for all but the most dedicated, in ideal regions. Muscadines, on the other hand, are highly disease and insect resistant as well as extremely productive.

Muscadines are plump, juicy, chewy, exceedingly sweet, and very high in sugar, with enough perfume and tang to make them interesting. The skins are usually thick and they have seeds, both of which picky modern folks dislike. They definitely do not crunch and pop like those strange, seedless, tart grocery store "grapes" do.

Muscadines are a wild USA native plant found from extreme southern KY south to Florida, and also in the warmer Mid-Atlantic areas. In most of the South they can be found growing wild hanging along trails and in forest clearings. Their *native advantage* is a huge one for the organic grower, because muscadines are very disease and insect resistant, especially compared to hybrid European and hybrid grapes such as 'Concord', 'Niagara', 'Thompson Seedless', etc., that often fail due to mildew, black rot, and so on, unless heavily sprayed. Muscadines are often grown successfully with no sprays at all. Some cultivars are more disease resistant than others, so choose carefully.

Cultivation

Muscadines are hardy to zones 7–9. In zones 7 they may experience occasional damage in an extreme winter event or the first winter or two. They may survive with protection and a warm microclimate in zone 6, but this is impractical for market growers. In colder zones of 5–6, experiment with Munson grapes (see Chapter 21). Muscadines are also not acclimated to dry or arid environments, preferring humidity.

As with all fruits, plant only where soil water drainage is good and avoid frost pockets and low-lying areas. Late frosts can burn back the sprouting shoots and reduce yield. Plant in full sun for highest yields, although they can tolerate some partial shade (this will strongly lower yields).

Have trellises built before planting. In spring only, plant *dormant* muscadine vines, either bare-rooted (preferable) or potted. Make sure plants are fully dormant (leafless). Stake with a 6' (1.8 m) tall bamboo cane at planting and continually tie the vine vertically to the cane as it grows. Keep well-irrigated and mulched to suppress weeds and conserve moisture. A healthy vine should reach the top wire of the trellis the first season, and make arms the 2nd season, from which the fruiting cordons grow. Substantial fruiting should start in year 3, peaking at around age 4–5. The vines are heavy feeders and should be fed with a high nitrogen, balanced fertilizer every spring and early summer. Do not over fertilize or fertilize after midsummer or in fall.

Training and pruning the vines is beyond the scope of this book, but

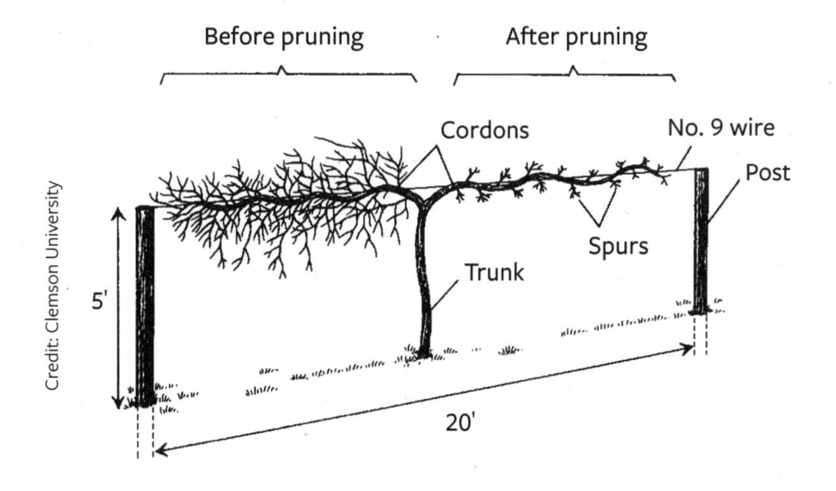

many southern US universities have easily accessible and free publications with very detailed instructions.[1] Also check those resources for detailed insect and disease control, which is usually minimally necessary. Neem oil is effective for both preventing leaf mildew and for insects, both of which should be minimal on muscadines if good cultural practices are followed, including providing good airflow.

The wild, spreading vines get big and the yields are very high, at maturity (years 4–5) around 50–80 lbs. (23–36 kg) per vine! Muscadines require rigorous summer and winter pruning to keep them in bounds and productive. Left unpruned, they rapidly become a rampant jungle of tangled vines.

It's recommended to plant the vines 12–20' (3.7–6.1 m) apart, so each of the two main arms can reach 6 or 10' respectively, each one going opposite directions on the trellis wire. They take about 2–3 years to come into production. Both of these factors make them more suited to market farmers with plenty of space and not urban growers. However, that urban micro-climate advantage mentioned earlier might pay off if you have a bit of land and plenty of time. The vines are extremely long lived, capable of living and producing for 50–100 years or more.

Irrigation

Muscadines are very resistant to drought once established and do not require irrigation except in severe drought conditions and sometimes during establishment if it gets hot and dry. Make sure to mulch new vines and any vines on sandy soil. Muscadines are not generally acclimated to dry summer or arid climates. Irrigation during initial fruit setting can increase yields but too much will water down the flavor. Overhead irrigation should not be used as it can encourage fungal diseases. Stop irrigation as soon as fruit is full sized and begins ripening. Utilize orchard tubing with two emitters per plant. Exact watering needs per season will depend on your region and soil type.

Pollination

Importantly, there are both *female* and *self-fertile* cultivars. The females *will not fruit* unless a self-fertile cultivar is nearby. The self-fertile cultivars *will* fruit without any pollination from another vine. Some of the more excellent cultivars are female, so pay attention to these details. If planting in rows it's recommended that the *self-fertile cultivars are planted as an entire end row*, then two rows of female cultivars, and then another row of self-fertile cultivars, and so on.[2] 'Fry' and 'Carlos' are good self-fertile pollinators. Vines are pollinated readily by wasps and native pollinators—protect them.

Some cultivars, such as 'Carlos', 'Lane', 'Scuppernong', 'Ison', and 'Sugargate' are more cold-hardy than others, yet they will *not* survive long-term in zone 6 without impractical winter protection. Cold hardy muscadines are sustainable in zone 7 only, despite what catalogs may say. Although, with special effort in protected microclimates, it's possible. I know a clever grower having success growing hardy cultivars in a microclimate as far north as Paoli, Indiana. Every winter he buries a low growing shoot to preserve it for when an inevitable winter kill event occurs, as it will in some years. That way he does not start back at zero, but digs up the shoot (protected under soil) and now has something to rapidly convert back into a trunk. He said he gets muscadine fruit about 3 out of every 4 years with this system, and likes growing 'Ison' best.

Pests and diseases

When well grown on a good site, and monoculture is avoided, and resistant cultivars are chosen, problems will be minimal to none. A few notable pests and diseases include:

Japanese beetles feed on muscadine foliage *much* less than European grape, a favorite target.

Leaf miners can be a minor issue, neem oil should help.

Raccoons can be an important issue and trapping may be necessary. They can climb vines and start eating the fruit even before it's fully ripe, in huge quantities, pounds per night. Left unchecked they might eat it all, and dislodge and ruin the rest!

Mildew on the leaves can be an issue if air circulation or pruning is poor, neem oil applied to the foliage as a foliar spray stops and also prevents it. This is an issue with our high tunnel grown vines, for which neem oil is very valuable and effective.

Marketing

Muscadines are a top choice for people in warmer climates looking to grow low-input, organic small fruits for marketing. In the South muscadines are a pretty easy sell. Wineries in the South are regularly popping up, focusing on making muscadine wine. There's a small market for muscadine grape juice and health products, both based around muscadine's high antioxidant levels. Muscadines were given a huge boost in the early 2000s by an enthusiastic Dr. Oz touting their strong anti-cancer antioxidants.[3] This led to a sudden, intense, and drastic increase in their demand and their notoriety overnight. Demand outpaced supply.

Grocery stores can move a lot of muscadines in late summer in the South, even big box stores often carry them. If you have a large amount of grapes and want to market them wholesale, talk to smaller local grocers or health food stores. Otherwise, farmers markets would be the ideal way to direct market them. Build a lot of hype weeks before you start to harvest via social media and at your local market. Make big, attractive piles of the fruit in bins, and have a basket scale and lots of bags and also pre-packed quart containers handy for quick sales. Larger growers might consider marketing bottled muscadine juice. Maybe even bring the easily-propagated muscadine vines in pots to sell. These propagate easily via layering and can sell for $5–20 each.

Online sales and shipping of muscadines are an eminent possibility. Their thick, durable skin comes in handy once again, as they easily handle shipment, especially in clamshell containers packaged safely. I've seen them going for as high as $11/qt. (0.95 l) online in 2021, and that's for conventionally grown! They currently list as selling in organic grocery stores for $8.99/lb. (0.45 kg) or more. If you got that price selling muscadines directly, and harvested 50 lbs. (23 kg) of marketable fruit from a single vine, that would be roughly $450 per vine/per season. That's vastly more than the typical market outlook for muscadine grapes grown and sold conventionally. Of course, use discretion and have realistic expectations.

Store freshly harvested muscadines in cold storage, or covered and at ambient temperatures for no more than a few days. The sweet fruit will draw yellow jackets and other animals and insects if left uncovered. Kept cold they keep a surprisingly long time in near-perfect condition, up to 4–6 weeks.

Cultivars

Many exist, but make sure to carefully research which cultivars can best handle your region's soil type and winter low temps (USDA hardiness zone). Focus on disease-resistance, productivity, and excellent quality, as well as regional adaptation and cold hardiness. Make sure to have self-fertile cultivars as well as females planted.

Self-Fertile Cultivars

Cowart: Excellent flavor, large clusters, and medium size fruit. Edible skin and strong production. Considered cold hardy and disease resistant. Ripens early season. Contains 17% sugar.
Ison: Self-fertile, black variety, 19% sugar. Excellent size and production. Excellent for wine production. Ripens early to mid-season.

Lane: A newer release from the University of Georgia. Produces medium to large black fruit. Ripens early during the first two weeks of August in south Georgia. Has a crisp skin and firm pulp which adheres to the skin. Very sweet with 19% sugar. Cold hardy to zone 7.

Dixie Red: Very productive, up to 17% sugar. The vine growth is very vigorous and cold hardy. Produces large clusters of reddish fruit. High yielding, edible skin, disease resistant. Ripens mid-season.

Carlos: Bronze grape, 16% sugar. Very vigorous, yielding large amounts of quality, medium-size fruit. Cold hardy and disease resistant. Considered good for white wine. Flavor is fruity and good with nice texture.

Nesbitt: Black fruited, productive, long season grape. Excellent for You-Pick and backyard use. Very tasty, large berries, and high in sugar. Disease resistant.

Noble: Black, self-fertile variety that produces medium size fruit with good quality and large clusters. Considered a good red wine grape. Wineries throughout Georgia, North Carolina, and Tennessee use 'Noble' to make red muscadine wine. Very vigorous grower, cold hardy, and disease resistant. High yield potential. Ripens mid-season. Contains 16% sugar.

Hall: Early to mid-season bronze variety. Almost the size of 'Fry'. Excellent flavor and very sweet.

Triumph: Produces medium size fruit with good quality and average production. Cold hardy and disease resistant. Ripens early to mid-season and contains 18% sugar.

Southern Jewel: Released by the University of Florida. High-yielding, disease-resistant, large black-fruited variety. This muscadine is unique, as it produces fruit in bunches of 6–12 berries that hold strong to the peduncle. Excellent flavor and a crisp texture with a palatable skin aiming for modern customer preferences.

Female Cultivars

Female cultivars require a self-fertile cultivar planted nearby in order to set fruit.

Fry: Considered the standard as it opened up muscadines to commercial production. Large greenish-bronze fruit with excellent quality and flavor. Edible skin. Vigorous, disease resistant, and cold hardy. Ripens mid to late season. Contains 21% sugar.

Pam: Produces bronze fruit, 21% sugar. Very heavy production and produces some of the largest clusters of all female varieties. Clusters can have 10–15 grapes.

Sugargate: One of the earliest ripening black muscadines. Excellent, exceptional flavor with large fruit and edible skin. Vigorous growers with good cold and disease resistance. Zone 7.

Supreme: Black muscadine, heavy producer with large clusters and dry scar. Has edible skin and is very disease resistant. Good choice for commercial vineyards. This variety may have a tendency to over-crop, so fruit thinning may be required. Contains 22% sugar—very high.

Summary

Delicious, extremely productive, very healthy, and easy to grow fruit. Requires stout trellising. The fruit is popular in areas where it grows. Exceptionally easy to grow organically due to the resiliency and disease-resistance of many cultivars, making them very low maintenance. Performs best in zones 7b–10. In zone 7 only plant the most cold-hardy cultivars. Make sure space requirements are adhered to as the vines are very vigorous. *Profit Potential*: Around $100–$450 per vine, maybe more, if direct retail marketing, and yields and subsequent sales are excellent, and at a high price point.

Urban Market Farming Rating—2/5: Trellising and space requirements limit its usefulness where space is at a premium. Would potentially do well in warm, humid, urban microclimates in zones 6b-7.

Rural Market Farming Rating—5/5: Easy to grow and harvest; without space restrictions this could be a great, low-input, low-maintenance niche crop in southern regions. Very productive and profitable. Great for You-Picks and farmers market sales.

Home Recommendations: Only plant muscadines on sturdy trellises designed for muscadines, and keep heavily pruned so they don't turn into a jungle. Don't plant grapes on chain link fences—they'll tangle through it and become a formidable woody mess. Go for the recommended cultivars for your region and make sure pollination needs are met (self-fertile planted with females). Go for disease-resistant cultivars only. Make sure to plant on a sunny, well-draining site only. Planting 3–4 vines will be plenty for most households.

Identifying Quality Stock: Lots of long, healthy roots. The 1 year old vines may be very thin but as long as the roots are good and the top is at least 1–3' long it's likely a good start. Avoid planting 4–5'+ (1.2–1.5 m) tall vines, it's not necessary.

16

Mulberries

Mulberries are a fruit of the future, one that I believe has massive unrealized potential in the USA and Canada. The fruit is nearly as good as a raspberry or blackberry, some would argue better, and the trees are naturally late blooming, cold hardy, extremely fast growing, resilient, disease and insect resistant, prolific fruit producers, and utterly reliable.

Before you doubt my mulberry raving, first, we're not talking about those weedy little off-flavored berries growing along your fence line behind your house in the ditch. We're talking improved cultivars. It's similar to comparing seedy, sour, wild blackberries to huge, marketable and productive 'Prime Ark Freedom' blackberries. Big difference. If you are skeptical or don't know much about improved mulberries you might be surprised to learn more.

Mulberries are in the family *Moraceae*, and are related to figs and also tropical breadfruit and jackfruit. In fact, if one closely examines and dissects a watermelon-sized jackfruit, one will find it resembles and is basically an enormous mulberry covered with a thick rind. All over the world, from India, Asia, the Middle East, Thailand, and much of Europe mulberries are prized as an excellent and versatile food source. And they are! From fresh eating to jams to dried fruit they have strong nutrition, intense flavors, and a lot to offer.

In former days of subsistence farming and during the Great Depression, mulberries were better appreciated. They were valued not only as

'Varaha' cultivar of *Morus rubra* mulberry selected by us, found in Louisville, KY. Note large, long berries.

human food but as free, nutritious livestock feed, especially for chickens, rabbits, and pigs. The sugar and protein-rich berries (as well as the nutritious foliage) raining from the trees made for easy feeding of farm animals and added much to their productivity, while also providing cooling shade, as well as erosion control on the farm. This concept is starting to be re-evaluated in the USA, especially in organic farming and silvopasture/permaculture circles. The cultivar 'Hicks' was apparently often used for this purpose, although many others would work just as well. The fruit is healthy human food: rich in iron, natural sugars, protein (unusually high for a berry), antioxidants, and fiber.

Species of mulberries

Mulberry fruit varies a lot, and so does their hardiness to cold, quality and quantity of fruit production, disease resistance, etc. *Be very careful that you are selecting a regionally adapted species as well as a regionally appropriate cultivar when selecting and buying mulberry trees.* There are a large number of species found around the globe but some of the most edible, available, and useful include the following:

Morus rubra: "Red" mulberry, native to the Eastern USA, from Massachusetts to northern Florida. In Massachusetts it is considered a threatened species.[1] In Kentucky it is fairly common outside of urban settings, along fences, in parks, and in forests. The long fruits vary in quality from decent to very delicious, with the best having a strong, tangy-sweet cherry flavor and firm texture. Trees in full sun are productive, sometimes heavily so, and the season is fairly short (2–3 weeks in early summer). It makes a great shade tree and has very large, lovely, heart shaped leaves. They grow very fast and establish rapidly, and also readily hybridize with *Morus alba*. They also make excellent rootstock for other *Morus* species. There is some concern that their genetic purity is being lost due to the encroachment of *Morus alba* into their territory, something which has already happened nationwide and is irreversible and is, in my opinion, not a crucial issue.

Morus alba: "White" mulberry, native to Asia, cold hardy to zones 4b–6, depending. *Morus alba* grows with gusto in most Eastern USA urban areas within its range, being viewed by many as a 'weed tree'. It's not unusual to see them growing and thriving along fence lines, behind garages, in ditches, just everywhere, and just raining berries for a few weeks every May or June, to the great enthusiasm of wild birds.

The fruit can ripen to either white, lavender, purple, or black. Berry size varies from quite small to large, up to 2 inches. The foliage is smaller and usually more ovate or sometimes fig-leaf shaped. Also sometimes referred to as *Morus tartarica* or Russian mulberry (possibly a subspecies). Berry flavor from wild trees tends to be mediocre to bland, but excellent wild specimens with abundant, delicious fruit can be found and many named cultivars with excellent quality and production exist. Several hybrids between *alba* and *rubra* exist as named cultivars.

Morus nigra: "Black" mulberry (native to Europe/Asia), cold hardy to zones 7 or 8. Ripens to dark purple/red or black with typically an outstanding sweet-tart flavor and smaller berries. Often black-fruited specimens of *Morus alba* are confused with and identified as *Morus nigra*. You will often see nursery and other listings for "black mulberries" but they often are referring to black-fruited *Morus alba*, so be careful when selecting. More cold-hardy *nigra* specimens are said to exist.

Morus macroura: The "Himalayan mulberry" is a common wild tree in Northern India and cropped there as well as in warmer parts of Asia, including tropical Thailand, where it is also a commercial fruit. Fruit quality is often outstanding and fruits are large, 1–4" (2.5–10 cm) long. The cold hardiness varies but the most cold-hardy specimens seem to be able to handle zone 7 and possibly 6 (we're trialing a number of them in zone 6 KY). 'King Shahtoot' and 'Pakistan' are cultivars of *macroura*. This species often has difficulty in areas with fluctuating or extreme winters as it buds out early, as soon as winter weather warms up. Flavor is sweet and often exceptional and they make an excellent dried fruit. May be best suited to warm, semi-arid regions, and for some cultivars, in the tropics.

Cultivation

Mulberry is extremely well suited to organic agriculture. They are one of the easiest and fastest to produce tree fruits, without much susceptibility to pests or diseases. The trees grow very fast, within 1–2 years from planting a whip (small tree) produces harvestable berries, and within 5–7 years full harvests can be expected. Mulberries bud and bloom late and escape most spring frosts. Even if hit with a frost they can re-bloom and set a partial crop.

Important to note is that mulberries come either male or female, with only females producing fruit, and males producing only (highly allergenic) pollen. Therefore, all named cultivars that produce fruit are female. Seedlings come either as male or female, and fruit quality will vary widely, with most *alba* seedlings producing mediocre to poor fruit quality, best suited to reforestation or rootstock.

The fruit *ripens early, well before most other berries are ready*, so they would not be easily overshadowed by the more popular blueberries or blackberries. They ripen in the Mid-Atlantic around the first or second week of June and are gone by early July (except everbearing types that continue to bear through August). Yields are very high. Mature trees can produce hundreds of pounds of berries per season.

Mulberries grow best in clay and loamy soils, yet they are very adaptable. Good soil drainage is highly recommended, although they can tolerate wetter sites if they drain. They like medium fertility, nothing too rich or else they grow too rank and flimsy, with branches wildly exploding out in all directions. Poor soil affects the berry quality and makes the berries smaller and more sour (lower in sugars, higher in acids). Healthy, fertile soils produce the juiciest and most flavorful fruit. The best I ever had were 'Illinois Everbearing' mulberries I planted on a silt loam soil in Louisville, KY. They like compost, mulch, and regular water availability. Rain can ruin the ripest berries but the trees *continually produce ripe berries every single day for a certain amount of time* (their harvest window) so rains, unless prolonged, are not usually very problematic. Although *drier regions are the most ideal for mulberry fruit cultivation*, many wetter regions like the Eastern USA (or Thailand for example) have strong potential for fresh fruit sales utilizing regionally-adapted cultivars such as 'Illinois Everbearing'.

For the market grower, it would be best to keep the trees *annually pruned down to shrub or bush size during the dormant season and again after fruiting is completed.* Mulberries handle this well and essentially start to

Note grafted mulberry fruit on the left, wild mulberry on the right. The difference in size is apparent, but so is the flavor, production, and overall quality drastically in favor of the grafted, named cultivar.

function as a shrub. This will drastically lower yields *per tree*, but what use are berries 30 feet (9 m) up a tree that you can't harvest unless you're a bird? You can keep them pruned to a reasonable height, 8–10' (2.4–3 m) or so, for easier harvest and maintenance. If not, the trees can reach quite large proportions, growing to 20, 30, 50 feet or even taller. They could be lopped down every winter to about 6–9' tall when dormant. Although not very well tested, it's plausible they could be kept even smaller and grown in a high-density setup. It would likely require cultivar testing to see which cultivars this succeeds with and which yield marketable quantities under such cultivation. This has been done in Asia.

Mulberries flower and fruit on *new spring growth* that sprouts mostly from one to two-year-old branches. So, maintaining plenty of smaller branches (very similar to *cordons* on grapes) is necessary. Trees can be planted about 8–10' apart and maintained as bushy shrubs, eventually forming a hedge-like arrangement. Another option is to pollard the trees low, around 2–3' or so, and allow a number of branches to sprout off the short trunk. In one year, those branches would be mature and would sprout fruiting shoots. After harvest, they could be lopped back to a reasonable height. Summer pruning controls vigor.

Credit: Internet unknown

Note the very closely spaced and pruned mulberries. The plants look to be only a few feet apart, and height is around 4 ft.

For a non-topped tree form, plant 20–25' (6.1–7.6 m) apart. At planting, prune off any growth below 2–3' (61–91 cm). Avoid bad crotch angles and prune off rubbing branches. The benefits of the tree form are potentially massive harvests of berries and large, low maintenance shade trees. However, harvesting the berries is impractical. Shaking the branches will dislodge the ripe berries onto tarps. This could work for mass processing of the fruit or animal feeding (in which case no tarp is necessary).

For market growing purposes, keeping plants pruned to a shrub size also facilitates netting them for bird protection, which will be necessary or at least recommended in most regions. Mulberry yield estimates are not known as there is little to no data available on this. However, from experience I would postulate that everbearing mulberries pruned as 8–9' (2.4–2.7 m) bushes will produce yields comparable to a similar square footage of blackberries or raspberries. Although pruning to shrub size reduces tree yield, yields would be high per square foot, probably higher than with larger trees and similar to better yields through high-density apple cultivation as opposed to tree-form apple production.

Left alone, many mulberries can reach quite large tree proportions. Wild *Morus alba* in a park in Danville, KY, likely 40+ years old.

Important to note is that many mulberries, especially tropical cultivars, are stimulated by summer pruning. Some growers prune off the tops of shoots, above a fruit set of ripening berries, to encourage faster ripening and large fruit size. Because mulberries fruit on new growth, whenever new growth occurs on one or two-year-old wood, fruiting is induced. Therefore, some growers have success with several rounds of summer pruning. By pruning back new summer growth by about ½ in the early summer, more sprouts can be forced on one and two-year-old growth. Those new sprouts will often flower and fruit. Some growers claim multiple harvests, up to 3 or 4 per summer, by summer pruning like this. This will also make the mulberry much bushier and shorter in stature. This is worth trialing as it could extend the harvest window which on many cultivars is pretty brief.

The trees are currently available from many fruit nurseries, either grown from *cuttings or grafted onto seedling mulberry* (usually *rubra* or *alba*) rootstock. They are fairly easy to propagate via greenwood cuttings and come into bearing within just a few years or less out of the nursery. Both grafted or cuttings-grown perform equally well, but cuttings-grown are likely preferable. With a cuttings-grown tree there is no need to worry about rootstock suckers, and if the trees die back to the ground any re-growth would be true-to-type to the cultivar. Choices will likely be very limited, as well as available cultivars. Grafted trees are a solid choice regardless.

Irrigation

Mulberries are very resistant to drought once established and do not require irrigation except in dry climates or in severe drought conditions. Irrigation is not needed in the Eastern USA once mulberries are established. Overhead irrigation should not be used as it can encourage fruit rot. If irrigating, utilize orchard tubing with 2–4 emitters per plant. Exact watering needs per season will depend on your region and soil type.

Potential for commercial production

Of all of these above species, the hybrids between *rubra* and *alba* seem to be the most commercially viable in areas with winters too cold to grow the superior *macroura* cultivars.

The better *rubra/alba* hybrids produce quality berries for an extended period (months instead of weeks), the berries often attain superior size (0.75–2" long, 2–5 cm), are firm, less soft and perishable, and have good to excellent flavor. The flavor is often reminiscent of figs, red raspberries, or blackberries and would be very attractive to most customers. I would postulate that anyone who likes fresh berries and tries some well-grown 'Illinois Everbearing' mulberries would become a fan *immediately*.

So, why aren't mulberries being widely grown and sold in the USA? It's the same reason that until very recently pawpaws (*Asimina triloba*) were not being marketed. Or, until the mid-20th Century blueberries were not being grown or sold widely. The answer: mulberries have simply not been developed or introduced as a commercial human food crop in the USA. Other regions of the world enjoy mulberries as a viable and sustainable commercial crop, including in many parts of Asia, India, and the Middle East. There's no actual impediment as to why they could not be a popular fruit crop in the USA or other similar temperate regions. They just need to be slightly developed/selected for commercial traits, trialed, marketed, and supported as a viable new crop.

One challenge is the *negative image* towards mulberries some carry, as a weedy unwanted tree that stains your driveway purple and makes birds leave purple droppings all over your car, or gets all over your kid's shoes. The name *mulberry* is not especially appealing either. However, these negative associations are minor and are not insurmountable. Many urbanites and younger people have no negative views towards mulberries; in fact, most urban people have no connotations or concept of mulberry at all. Appearing at the farmers market and grocery stores packed nicely in clamshells, and tasting great, they'd rapidly catch on and likely become a hit.

Harvesting

Most quality dark-fruited cultivars *can be harvested when partially ripe (red)* and have a somewhat more tart, raspberry-like, flavor and firmer texture. When purple/black they are juicier, slightly softer and have a flavor all their own but reminiscent of figs, cherries, and blackberries, usually with a nice tangy zing. When picked, ripe mulberries retain a tiny green stem which is edible and usually unnoticeable. The berries can be shaken from the trees, but unless they are all going to be processed it's usually better to harvest by hand. Shaking the trees dislodges plenty of leaves, unripe berries, twigs, bugs, and other duff that will need to be cleaned out. Perhaps professional berry cleaning equipment would perform that task.

The berries are fairly soft, and firmness varies from one cultivar to the next, so very careful harvesting is necessary, as well as cultivar selection. Don't stack berries more than a few layers high, and keep them out of the hot sun. They will stain your hands as you pick. The stain fades within a day or so. Higher quality cultivars produce firmer fruit less likely to leak juice and better for marketing; 'Pakistan', 'World's Best', and 'Illinois Everbearing' are three examples.

Pests/Diseases/Challenges

These are usually minimally impactful, but do exist. Likely the main disease of concern is the fungal infection known as *popcorn disease* (*Ciboria carunculoides*).[2] It is only known to infect mulberry fruit in southern US states, mostly south of Tennessee. I've never seen it in Kentucky but have heard of one report of it in the state. It makes the individual carpels (the tiny individual seed-containing berries comprising the aggregate fruit we call a "berry") swell up and turn tannish in color, resembling popped popcorn. Obviously, this ruins the fruit, usually most or all of it and makes mulberries in the South a questionable proposition. However, there are resistant, high-quality cultivars. See the Mulberry Cultivars section for more information. Bordeaux mixture or other organic fungicides can likely be used on immature, green berries to prevent infection, organically. Popcorn disease may be mostly a problem affecting *Morus alba*.

Popcorn disease of mulberry fruit.

Insects

Stinkbugs The most notable insect issue is likely stinkbugs. Various species feed on mulberry fruit. They often leave an unpleasant off-flavor of *stinkbug* on the berries. Usually, their population is minimal early in the season when mulberries ripen.

Ants can take interest and consume a lot of berries, and if they are problematic I would suggest keeping the grass cut very low and putting a Tanglefoot® sticky band around the trunks.[3] To apply, you simply wrap the trunk in a layer or two of duct tape, and smear the Tanglefoot® on the duct tape—not onto the bare trunk! If the product touches the trunk, it can damage the cambium of the tree and can kill it. It lasts a few months at best, long enough to get through harvest. The berries ripen before SWD becomes an issue.

Aphids can be problematic but are usually isolated to stressed plants or individual twigs. A strong soap spray or strong jets of water from a hose would take care of aphid problems. If you observe large ants frenetically running on the trees or twigs, this is usually a sign of aphids being farmed for honeydew (by the large ants) on your plants in great numbers. Utilize Tanglefoot® and release ladybugs and lacewings. Spray organic pesticides only before fruit ripens or after harvest is complete. There are no registered pesticides for mulberry trees being grown for fruit production.

Tent caterpillars (similar to silkworms) can take up residence in early to mid-summer and their presence is very obvious. They form a colony of dozens or hundreds of caterpillars, all living in a matrix of thick silky webbing on the ends of a branch. They rapidly decimate the foliage and twigs and can defoliate large sections of a tree if not checked. It's easy to tear them out and crush the webs, caterpillars and all. They can also be fed to chickens, who relish them. Old-timers would light kerosene torches on long poles and torch them, tents and all. Old-timers were badass.

In some regions, including Kentucky, ripe mulberries, if observed very closely, will often, but not always, reveal very tiny thrip-like insects crawling on them. These are harmless and could be the larvae of various species of tiny insects. Most customers would never notice them, and they do *not* appear to affect fruit quality. Apparently, you can wash them off but this would certainly make the fruit perish very quickly. It's possible that refrigeration of the berries would likely "deactivate" the tiny insects quickly, similar to SWD larvae. Again, most customers would likely never even notice them, but take note. In drier climates these insects are likely not present.

Animals

Birds, such as robins and cedar waxwings, certainly take interest and could make off with most of the harvest. In fact, they prefer the protein-rich mulberries to other less nutritious fruits. Some people plant mulberries mostly as a *trap-crop* from their more prized fruits like cherries. This is where netting and other bird protection becomes necessary, and therefore keeping the trees pruned to shrub size makes even more practical sense.

Deer relish mulberry foliage! It's a favorite browse. So, deer cages or deer fencing will be required outside of urban areas.

Other challenges

Late frosts can diminish the crop, but the trees will usually produce a small second flush of blooms if the first set of blooms is lost, making at least a partial harvest possible. However, make sure to plant the trees in *higher locations* and not in low-lying "frost pockets." You could utilize all night *mist* or smudge pots around mulberries to protect expanded buds or shoots in case of late frosts.

Marketing

Mulberries are a gem in the rough, just waiting for enterprising, progressive growers to capitalize on them, and an ambitious university to trial and develop them. The cultivars are there, what we need are production strategies and systems. Everbearing mulberries would fit in perfectly to a mixed-farm growing small fruits and other crops, as well as silvopasture and agroforestry.

Once customers were educated and tasted the delicious berries they would be hooked. In recent years I've observed a small market developing in the western US states, with market growers packaging and selling excellent mulberries just like any other berry, hyping them on social media, and customers are enthusiastically buying. Often the cultivar 'Pakistan' is being grown and marketed in these areas, and it is well suited to the Southwest.

With the ease of growth and high yields, the growers are enthusiastic too. For marketing, ½ pint, pint, and quart clamshells would all work fine. Make sure not to stack the berries too high and crush them. Mulberries are slightly more perishable than other berries due to their high water content, and should be chilled as soon as possible after harvest. In refrigeration, firmer cultivars might keep about 3–4 days. Sell ASAP.

Value-added potential is high. Dark-fruited cultivars make luscious, rich jams and good jelly. Desserts made with mulberries have great texture and flavor; the tiny edible seeds also add a slight nutty crunch. For thousands of years they have been dried like raisins (only white or yellow cultivars are typically dried and are non-staining). Certainly, other possibilities exist: juices, smoothies, ice cream, wines, fruit leather, etc. Powdered they could become the next health food "superfood" fad. If goji berries made it as far as they have, certainly the vastly superior flavored, more productive, and much larger mulberry could do it too. Other uses include dried leaves for a healthful tea, or dehydrated and marketed as food for wild animals, pets, or livestock. They have solid and very diverse marketing potentials.

Farmers markets would likely be the best place to start marketing, as the "public education" and free sample aspects of marketing will be important to get this crop popularized. Local bakeries or ice cream shops could possibly be convinced to try them out. Overall, they would be an easy sell if large, high-quality berries were packaged attractively and free samples were given out to create interest.

The future of mulberries

Utilizing superior mulberry genetics such as 'Illinois Everbearing', 'Chaing Mai #60', 'Thai Dwarf', 'Pakistan', 'King Shahtoot', 'World's Best', etc., mulberry breeding and hybridizing could enhance the fruit size, flavor and picking season of the berries.[4] Also, many superior cultivars exist in Europe and Asia, they just have to be identified and obtained.

Through specialized training and pruning to keep the trees in bounds and able to be netted for birds, or even through dwarfing rootstocks, yields can be guaranteed. A natural Asian dwarf species *Morus kagayamae*, has potential for this, or perhaps using dwarf cultivars of *alba* as rootstocks or interstems.

Marketing of delicious mulberry products and value-added goods would increase demand substantially. We need to simply follow Asia's example and copy their systems with this. Mulberries can often be substituted for blackberries or raspberries in recipes—to good, slightly different, effect. Perhaps the fruit should even be renamed. How about: jam berries, yum berries? Or, we could look to other languages: in Portuguese: *morango*. Italian: *gelso*. In my opinion, *morango berries* or *jam berries* sounds a lot more appealing than *mulberries*.

The fact that they require no pesticides or fungicides to produce good crops is a winning point right there. This fruit is just patiently waiting for a university agriculture program to pick it up and roll with it, and turn it into something much more appreciated and valuable to humanity than it is currently, at least in North America.[5] But until then, we already have some great genetics to work with and a rapidly growing interest in growing the fruit for home and market. Any adventurous market grower could begin immediately benefiting by adding this promising crop to their operation. Likely production will be focused in drier, more arid western areas but certainly has much potential in the eastern USA as well, and that should be developed.

Recommended cultivars

Illinois Everbearing: Excellent tangy-sweet grape-like flavor and heavy production. Produces ripe berries for about 6–8+ weeks. Berries are fairly long (1–1.5", 2.5–4 cm) and plump, ripening to purple-black. For now, it remains the most marketable and superior hardy cultivar for growing in humid, cold-winter regions like the Eastern US. Very vigorous, stout, and cold hardy. *Rubra/alba* hybrid. Zones 4–9.

Hicks: Old cultivar supposedly similar to 'Illinois Everbearing'. Heavy yields with large, sweet berries. Formerly used for silvopasture use, for which it is well suited. From NC. *Morus rubra, possible hybrid.* Zones 5–9.

Silk Hope: Said to be similar to 'Illinois Everbearing', but resistant to "popcorn disease" found in the South. From North Carolina. Delicious berries. *Morus rubra*, possible hybrid. Zones 6–9.

Miss Kim: Plump, tasty, black berries, good production. Originates in Georgia. May have popcorn disease resistance. *Morus alba*. Zones 6–9.

Beautiful Day: Very sweet but very small, pure white berries. No staining. Berries too small to market, for home use only. From Florida. *Morus alba*. Zones 6–9.

Madhava: Tasty, plump, soft black berries in abundance. Very good flavor. Makes a great backyard specimen. Zones 6–9. *Morus alba*. Selected by Peaceful Heritage Nursery. Found in Louisville, KY.

Varaha: Long, reddish-purple berries in abundance. Very good flavor. Trees are vigorous. Heavy production. Selected by Peaceful Heritage Nursery. Found in Louisville, KY. *Morus rubra*, possibly hybrid.

Collier: Berry quality similar to 'Illinois Everbearing'. Very cold hardy heirloom selection. Proven reliable and good quality. Zones 4–9. *Morus rubra/alba*?

Kokuso: Korean species. Very cold hardy with heavy berry production. Berries are firm bordering on dry. Mild, sweet flavor, sometimes bland. Zones 4–9. *Morus latifolia*. Productive and cold hardy in marginal areas but flavor is lacking. Semi dwarf tree stays about 15 × 15' (4.5 × 4.5 m). Makes a good processing berry when sweetened.

Gerardi (Geraldi) Dwarf: Fruit quality similar to 'Illinois Everbearing'. Grows very slowly and stays permanently in the 10–12 ft tall range, but produces fruit rapidly. Heavy production of long, quality berries and has a long season. Zones 6b–9, possibly colder. Rubra/alba hybrid. Other dwarf mulberry cultivars exist but most produce small, mediocre berries. For a comprehensive cultivar list, check out www.growingmulberry.org

Vratza 24 and Vratza 18: Large, black, tasty berries with a long 2–3 month season. European *alba* cultivars.

Warm climate mulberries

King Shahtoot (Saharanpur Local, Australian Green): Very long (3–4", 7.5–10 cm) yellow-white berries of exceptionally sweet, superb quality.

Heavy production. From India. *Morus macroura*. Zones 8–10, best suited to drier locations.

Maui: Plump black berries with good flavor and production. Twigs are greenish. Comes from wet and humid, tropical Maui, Hawai'i, yet appears in our trials cold hardy to at least zone 7, maybe 6. *Morus alba*.

Pakistan: Popular cultivar makes long, 2–4" long reddish-purple berries of excellent quality; drier and firmer than many others with good flavor. *Best adapted to the Western US* and mild winter areas only. Will begin budding during occasional winter warm-up periods, thus getting very damaged or dying when it freezes again, making it unsuitable in most eastern US locations with fluctuating winter temps. Excellent marketing potential for the fruit. *Morus macroura*.

World's Best: Heavy and early production of plump, sweet, mild-flavored berries. Blooms early, possibly similar to Pakistan, so may not be well suited to upper Southeast or areas with fluctuating winter temperatures, but it is currently under trial in many areas, with mixed results. From Thailand. Zones 7–10, possibly colder. *Morus alba?*

Chaing Mai #60: Thai cultivar grown commercially in Thailand. Difficult to acquire. Heavy production of excellent berries. *Morus macroura*. Zones 6–9.

Noir de Spain: Excellent flavored black mulberry. *Morus nigra*. Zones 7–9.

Chelsea King-James: A highly regarded *nigra* cultivar with excellent fruit, from London. *Morus nigra*. Zones 7–9.

Thai Dwarf: An outstanding dwarf variety with very large, firm, drier black berries with very good mild flavor with zing. Extremely productive. Cold hardiness undetermined. *Morus macroura*? Zones 7–9.

Summary

Mulberries have great untapped potential as a commercial fruit and small market fruit, especially in drier, mild winter regions where the best cultivars thrive. Low input and very easy to grow. Keep the trees heavily summer-pruned to shrub size and experiment with pruning to stimulate multiple fruit sets. *Careful selection of species and specific cultivars suited to your climate region will ensure quality fruit and survivability.* Birds sometimes pose a significant challenge and so netting could be necessary. Other than that, almost pest-free. Vast potential as a new crop. *Profit potential*: Semi-dwarf trees: $50–150 per tree in humid regions, and in ideal (drier) climates $200–300 per tree (estimates). Bushes: $20–30 per bush, maybe more in ideal climates.

Urban Market Farming Rating—2/5: Fairly space efficient per pound of fruit produced but the shrubs can get large and could shade other crops (could be an advantage) and make extensive root systems. Large, quality berries could do very well at market in clamshells. Ethnic populations (Indian/Middle Eastern/Asian) might be enthusiastic about mulberry fruit in urban places. May be best as a rural crop brought into the city to market.

Rural Market Farming Rating—5/5: Easy to grow and without space restrictions this is an excellent niche crop to look into. Again, cultivar selection and tree training is crucial. Much, much potential especially in drier regions.

Home Recommendations: Choose 'Illinois Everbearing' and 'Gerardi Dwarf' in humid regions, or 'Pakistan' and 'King Shahtoot' in the Southwest and California. Keep them cut back very hard unless you want a fairly large tree. Plant in full sun. Dwarf cultivars are worth experimenting with but careful selection is important.

Identifying Quality Stock: Lots of vibrant yellow roots and a pencil-thick diameter trunk at least 1–3' tall. They grow extremely fast so a large plant is not needed to start. Cuttings propagated are more desirable than grafted, but both work quite well.

17

Gooseberries

Gooseberries are a tasty and healthful fruit that's very popular in Europe, though not very well appreciated in the USA. With their ease of growth and productive nature they could be utilized much more in the future. Their name does not contribute to their appeal, however, at least in my opinion.

Gooseberry is an unassuming, low-growing, often spiny shrub with little goose-foot shaped leaves. The round, pea size berries with vertical stripes resemble spherical little watermelons. They are crisp, sweet-tart, and taste similar to kiwi. The familiar kiwi fruit of commerce was previously named "Chinese gooseberry," because it tasted like, well, gooseberry.

Its species *Ribes grossularia* has for centuries, or millennia, been a cultivated crop in Europe, and is quite popular in most cooler areas still, including England, Germany, and Poland, as well as Russia. In the USA, they are most popular in New England and other colder, northern areas. The fruits ripen to shades of green, yellow, red, pink, and purple. There is a native American species (*Ribes hirtellum*) and cultivars available, and they can be found growing wild in the eastern USA. The native gooseberries I found in Missouri once were very small and tart.

One thing that has held back gooseberries in a big way is that the plants were once federally illegal across much of the US. The reason was that gooseberries are a factor in the lifecycle of a devastating white pine tree disease called white pine blister rust. It first infects gooseberry foliage, doing minimal damage to the gooseberry, and then finishes its life cycle on white pines, where it often kills the trees. Due to white pines being a strong economic factor for many states at the time, the US made it illegal to cultivate gooseberries in those states. That ban is now lifted in most (but not all) places, so check your state and county laws.

There are also a number of similar hybrid species, including various **jostaberries** (complex hybrids of *Ribes divericatum* and gooseberries, and also **Worcester berries** (*Ribes divericatum*). These fruits are best utilized for jams and will likely have little to no appeal in the USA, but may still have potential in Europe. Personally, I have never had success with jostaberry. There are different strains and it succeeds some places, being popular in Germany.

Resistance to extreme cold, diseases, insects, late frosts, and inclement weather make gooseberries a 21st Century fruit waiting to happen in the USA. Early season fruit, very quick to marketable production, and taking up little space further enhance their draw. As I've grown them over the years, I appreciate them more and more. The luscious jam made from them has an excellent kiwi-strawberry flavor—if sweetened heavily with sugar! Folks I've introduced ripe fruits to all liked them. One person compared them very favorably to sour fruit candy. Kentucky State University (KSU) did a study and trial plantings of gooseberries and currants in the early 2000s, concluding that gooseberries are easily productive, hardy, and can be grown and marketed in the upper South and Midwest.[1]

Cultivation

While generally easy to grow, one of the limiting factors with gooseberries is their intolerance to extreme heat and humidity. This makes growing them in the hotter parts of the south and Gulf areas unrealistic. You will need to ascertain whether your region's heat and humidity levels are acceptable for gooseberry culture. They thrive all across the Midwest, upper South, and New England zones 3–7, where summers are mild to moderately hot and winters extreme to moderate, and do well in many parts of Canada. In hotter zones, planting them where they receive shade part of the day in summer, especially in the afternoon or late afternoon, seems to be helpful.

The blooms are very resistant to frosts, making them hardy to late spring frosts. I've seen them in full bloom with fruit, survive 28°F (–2°C) with no apparent damage to the blooms or even very small immature fruits, which is highly unusual among edible fruiting plants.

At planting, choose the best looking 4–6 branches, remove the weaker ones, and prune the remaining ones down to 6–8 inches, removing flowers the first year of planting to help them establish. Do not fertilize the same year of planting. High soil organic matter helps gooseberries thrive. Plants should be set 3–4' (0.9–1.2 m) apart in the row, with rows 6–8' apart.

Pruning involves removing old canes over 3 years old, at their base, and allowing new ones to grow for 3 seasons before removing them. You could also remove ⅓ of the largest canes each year. Remove crossing and drooping branches. Some cultivars can become "weepy" and this downward growth and any branches at ground level need to be pruned back. Alternatively, these can also be shallowly buried to root and easily create new plants (layering). These are a valuable commodity.

Like currants, gooseberries need to be rejuvenated every 4–6 years by choosing new branches arising from the ground or low healthy branches. These will become the new main branches. Remove all older growth and focus on leaving 1-year old branches well-spaced.

Irrigation

Gooseberries are resistant to drought once established and do not require irrigation except in dry climates or in drought conditions. Irrigation during initial fruit setting can increase yields but too much will water down the flavor and could make fruit split. Overhead irrigation should not be used as it can encourage foliage mildew. Stop irrigation as soon as fruit starts coloring up. Utilize orchard tubing with 2 emitters per plant. Exact watering needs per season will depend on your region and soil type.

Pests and Diseases

Good airflow via proper plant spacing and pruning is necessary to prevent foliage **mildew**, the most serious disease issue. Mildew-resistant cultivars are highly recommended. Neem oil sprays after blooming could possibly help.

As mentioned earlier, **pine blister rust** is a serious disease of pines, but the effects on the foliage of gooseberries and currants is not anything to worry about. Check local regulations on gooseberry and currant bans.

Japanese beetles are drawn to feeding on the foliage of gooseberries. Thankfully, they tend to show up in early summer after the gooseberry harvest is already complete and the plants have already put on much of their summer growth. Pyrethrums provide rapid knock-down of Japanese beetles, and are safe to spray after berries are harvested. Some cultivars seem less prone.

The most serious pest is considered the **imported currant worm**[2] *Nematis ribesii. If it gets to be severe, bushes can be sprayed with Surround® after blooming, covering the small fruitlets, and also pyrethrins and spinosad can be effective as long as honeybees are gone and flowering is well over.*

Sawflies (*Janus integer*), considered a major pest in Europe, girdle the stems and bore into the canes. I have seen these on our farm in KY, but nothing serious. Removing wilted shoots and discarding/burying provides control.

Overall, however, gooseberries are remarkably free from serious diseases and insects when resistant cultivars are planted in suitable locations with good cultural care.

Harvesting

Gooseberries ripen sequentially on the bushes, with most of the harvest being ready within about a two-week window. Carefully harvest only ripe berries, as underripe berries and ripe berries mingle freely next to each other on the branches. This can be a little tedious. Cultivars ripen to various different colors and shades, so understand the cultivars you are growing and ripeness indicators. Overly ripe berries become mushy in texture and fall from the bushes. These are often fine for processing or home use.

Commercial varieties tend to ripen at about the same time on the bush, making harvest easier or mechanically suited.

The best way to harvest is to wait until most of the berries are at or near peak ripeness, and, wearing thick gloves, clench a branch in your fist and gently drag your hand down the branch, stripping the berries off into a harvest tub below. A berry rake might work well for this fruit. Pick out leaves, duff, and small, underripe berries. Chill the berries. They can keep in cold storage about a week or so, maybe longer.

'Invicta' gooseberry makes lots of high-quality berries that ripen to greenish yellow.

Marketing

I only recommend marketing gooseberries if they are in demand locally or you are planning to go on an educational campaign. Are gooseberries a cultural or culinary "thing" where you live? Are there lots of European/Russian local immigrants? If so, you already have a customer base.

You can also survey your customers at the farmers market about gooseberries. Are other people selling gooseberries? Do you get asked about them? If so, it sounds like gooseberries could be a winner. If local folks don't even know what a gooseberry is, it's probably best to not invest in growing gooseberries for local sales. Again, unless you're ready to start an educational crusade and risk trying something different. You could also add them to your You-Pick operation, planting them, say, underneath your main trees or crops, and "broadcasting" that you have gooseberries also. Sometimes when word gets out in an ethnic community that people have something highly prized and hard to source, the response can be surprisingly intense.

Fresh berries can be marketed in ½ pint clamshells, pint clamshells, and quarts. The berries are fairly firm and dense and so will not easily crush or leak juice. They keep well in refrigeration, for at least a week or more, so picking could be gradual throughout the week. Ascertaining the correct color of ripeness can be tricky, so make sure you are picking the berries when fully (but not overly) ripe. Your tastebuds can help educate you on when they taste the best.

As mentioned, they make excellent jam, and so could be marketed also in a number of value-added goods. Note: we have found raw honey alone added to already cooked, unsweetened gooseberry jam will sweeten gooseberry jam, but it still retains a (delicious) sour bite (according to Ayurveda you should never cook honey). If you want it 'super sweet,' cook with some amount of sugar. Local bakeries and sweet shops could be interested, especially if gooseberries are already a cultural thing.

It should be mentioned that German, Polish, English, Romanian, and other European people, as well as Russian and Ukrainian populations would be *very interested* in fresh or processed gooseberry products, and purchasing the bushes themselves. If you live where these populations exist, local gooseberries could be a hit. Ethnic grocery stores catering to these populations would likely be very interested as well. Maybe utilize Google translator and bring photos and pricing info to seal a deal.

Recommended cultivars

Many are available, this is a small modern sampling of high-quality cultivars. Choose disease-resistant ones adapted to your climate that abundantly produce large, quality berries.

Black Velvet: High quality, nearly black berries said to have a blueberry-like flavor. Vigorous and disease resistant. Good for fresh sales.

Jahn's Prairie: Very disease resistant with large, flavorful reddish-pink berries. Yields comparable to 'Invicta'. Fruit quality is comparable to high quality European varieties. Erect growth habit with some sprawling branches. Very productive. Selected in Canada.

Jeanne: Suitable for commercial production due to very high yields, quality, and disease resistance. Thin skin and excellent flavor. *Blooms and ripens later than most*, with fruit ripening around mid-July, providing season extension possibilities.

Poorman: American cultivar with excellent quality red fruit.

Invicta: A newer cultivar with high yields of flavorful medium-sized, sweet green/pink berries. Somewhat spiny. Very high-quality fruit and strong production. Erect growing habit up to 4–5 ft tall makes harvesting easier. It's somewhat a Japanese beetle magnet. Not as disease resistant as some others, prone to losing foliage in KY by August-September.

Hinnomaki Yellow: A flavorful European/American hybrid cultivar said to have a flavor reminiscent of apricot. From Finland.

Hinnomaki Red: A hardy European/American hybrid cultivar. Produces a sweet/sour, marble-sized reddish berry in abundance. Ripens early. Disease and insect resistant. Healthy, spotless foliage. From Finland.

Captivator: European/American hybrid. Sweet, tear-drop shaped red berries up to 1" in large clusters. Very productive and less spiny.

Oregon Champ (Oregon Champion): Very vigorous, often reaching up to 5' tall. Very productive of large, green tart fruit, perfect for processing.

Tixia®: Newer trademarked cultivar makes large, high quality red berries in profusion.

Welcome: Sweet, large ½" pink-red ripe berries. Less thorns than most. Excellent flavor. Hardy to −40°F. Better shade tolerance than other gooseberries. Hardy to zone 3.

Summary

Delicious small berries on easy to grow bushes. Needs a region or location that is not arid or extremely hot in summer or overly mild in winter (USDA zones 3–7). Demand will be low to zero in many US locations, but would be a hit in others, so choose this crop with careful consideration as to local market. Great for value-added goods like jams. Good for You-Picks. *Profit potential*: $15–50 per bush, maybe more in a strong market.

Urban Market Farming Rating—4/5: Fairly space efficient per pound of fruit produced. If demand is high, this could be a good urban farming crop due to heavy production and small space requirements. Bushes are quite compact and do not require trellising. They also can tolerate light shade.

Rural Market Farming Rating—4/5: Easy to grow and with good customer demand this could be a solid niche crop to look into, especially in northern and prairie regions. Some cultivars can be grown under fruit and nut trees. Would be 5/5 in areas with high demand.

Home Recommendations: In warm summer areas plant where the bushes receive afternoon shade. Trial a few cultivars to see what works best and what tastes best to you. Just 3–6 bushes should supply most households abundantly. Harvest when ripe so they don't all drop or get eaten by birds and bugs.

18

Currants

Currants are an extremely hardy, popular, and iconic small fruit of Europe. Formerly more popular in the USA, yet currently not that well known or consumed in many regions. Similar in most ways to gooseberry, they are again culturally sought out by people living in or from colder areas of Europe, Ukraine, and Russia.

Currants should only be considered if you have (or are willing to create) local demand for the fruit. If the demand is there, these will be an easy sell.

Currants come in pink/red, white, as well as black fruited varieties, each with distinct flavor and appeal. Black currants are strong flavored and musky. The delicious flavor is acquired. They excel as a processing fruit and are extremely productive. Black currant juice and jam is extremely popular in Europe. Red currants are more tangy while white currants are the most mild and sweet, although all have a slight musky or slightly astringent edge. The small, soft seeds are edible.

Cultivation

Currants need very similar growing and climatic conditions as gooseberries, to which they're related. They grow best in areas with cold winters and mild to moderate summers without intense heat and humidity. These berries grow best in prepared 30" (76 cm) beds.

Start with high quality, disease-free, bare-root stock, preferably 1–2' tall plants. Plants need not be large as they grow rapidly and start

Delicious black currants are very prolific when the bushes are well-grown.

producing in 1–2 years from planting, or up to 3 years on some slower cultivars. Plant in early to mid-spring about 3' apart. Some vigorous cultivars may require more space, so research spacing on your specific cultivar. Planting too close will lower yields and encourage mildew.

Red/Pink/White currants: Plant the bushes 3–4' (0.9–1.2 m) apart in the row and 6–8' between rows. At planting, cut them back to 6–8". When dormant-pruning mature plants, remove all 3-year-old and older branches and canes. These currants produce mostly on spurs on 1 to 3-year-old branches.

Black currants: Plant them 4–5' (1.2–1.5 m) apart in the row and 7–9' (2.1–2.7 m) between rows. At planting, cut them back to 2 buds near the base. At dormant pruning, choose 5–6 one year old stems and remove all other stems older than 2 years. Black currants grow very robustly and can get quite large, about 4–5' tall and wide, and produce large harvests. Black currants bear on 1 and 2-year-old wood.

Importantly, black currants can grow in the shade of other tree crops, such as Chinese chestnuts, apple/pear, or pawpaw. This makes them a useful addition to orchards and You-Picks. Progressive farms in the Midwest are experimenting with these mixed crops and some success is being had.[1]

Currants can put out 15–20 lbs. of fruit each. They can be labor-intensive to harvest and some cultivars are quite soft, especially black currants. Research cultivars very carefully to make sure you are getting a suitable one for your region and plans.

Interestingly, in Europe currants are being grafted to special rootstock to convert them into taller shrubs, and also being grown in high tunnels. This is not available in the USA.

Irrigation

Currants are somewhat resistant to drought once established and do not require irrigation except in dry climates or in drought conditions. Irrigation during initial fruit setting can increase yields but too much will water down the flavor and could make fruit split. Overhead irrigation should not be used as it can encourage foliage mildew. Stop irrigation as soon as fruit starts coloring up. Utilize orchard tubing with 2 emitters per plant or use drip tape. Exact watering needs per season will depend on your region and soil type.

Pests and diseases

Insects are the same as for gooseberries. Diseases are similar too but differ slightly.[2]

Anthracnose: is a fungal infection appearing first as numerous dark-brown to black dots scattered randomly over one or both surfaces of foliage. Anthracnose can manifest at any time during the growing season. The spots soon enlarge and affected leaves turn yellow and drop. This weakens the plant, reduces its vigor and productivity, and results in smaller fruit of lower quality.

Leaf Spot: sometimes called *septoria* leaf spot, this leaf spot can be distinguished from that caused by anthracnose. The spots typically appear on the foliage in June, and resemble anthracnose. Spots soon enlarge and the central area becomes lighter in color with a brown border. Tiny black specks then appear scattered over the surface of each spot. These black specks do not appear with anthracnose. The diseased leaves, especially on currants, soon yellow and drop.

Powdery Mildew: The American species of mildew is the most serious pest of gooseberries, but is seldom on currants. White, powdery patches first on the lower (shadier) foliage of the bush, attacking the leaves, shoots, and also the berries. As it progresses, the surface of these parts becomes covered with a whitish powdery fungal growth. Strong mildew infections will cause stunting and premature dropping of the foliage, lowering fruit production and weakening the plants. Choose mildew-resistant cultivars only.

Birds can take interest in red currants and they will likely require netting. Black and white currants are less bothered by birds. Other animals are usually not an issue.

Harvesting

Harvest fruit when fully colored and ripe. Currants, especially blacks, are very soft when super ripe and are slow to harvest and easily damaged. They require careful handling and no stacking. Harvest directly into clamshells. Chill immediately after harvest and direct sell within 2–4 days. If processing the fruit, you can harvest whole bushes similar to gooseberries. Currants have no thorns. A berry rake should work well to help harvest this fruit.

Marketing

Generally, the same rules and standards apply for currants as for gooseberries. See the gooseberry section for marketing advice. Currants have a lot of potential in Northern and Midwestern regions where/if customer demand exists. European people love this fruit and demand is strong. Could fit in well in You-Pick farms and orchards. Should be much more popular. Focus on growing only those most marketable in your area (black, red, pink, or white).

Ethnic European populations would be very interested in this healthful fruit, especially black currants, and will come to harvest the fruit if informed. Spread the word. Black currants could be a real score if you have the market demand.

Cultivars

Black

The 'Ben' series are named after the mountains of Scotland and produce extremely high-quality fruit.

Ben Sarek: Compact growth habit excellent for high density and You-Pick your own farms. Bears consistently and heavily and may need some support. Susceptible to powdery mildew.

Ben More: A very productive cultivar with excellent quality.

Ben Lomand: A compact growing bush attaining 5' tall. Strong pungent flavor. Mildew resistant.

Titania: A newer cultivar with mildew resistance and high yields of large, high-quality berries. Very vigorous; plants get up to 6' and are said to be immune to blister rust. Self-fertile.

Red

Red Lake: Most popular red cultivar in the USA. Long fruit stems and easy to pick. Long bearing season. Self-pollinating. From University of Massachusetts. Susceptible to powdery mildew. Zones 2–8.

Rovada: Abundant production with excellent quality. Late flowering, very reliable, vigorous, and mildew resistant. Zones 3–7.

Jonkheer van Tets: Large clusters with heavy production and good quality. Excellent for espalier. Mildew and aphid resistant. Cannot handle intense summer heat. From Holland. Zones 3–8.

Honeyqueen: Long harvest period of soft, delicious fruit. Resistant to mildew.

White

Pink Champagne: Compact bushes produce large clusters of fruit. Good flavor. Self-pollinating. Mildew resistant. Ripens in July. Zones 3–8.

White Imperial: Excellent flavor and quality. Berries are translucent and small-medium in size. Zones 3–8. Introduced in 1895.

Blanca/Blanka: Very high sugar content. Good shelf life for market. Late bloom time. Self-pollinating. Zones 3–7. Mildew resistant. From Sweden.

Summary

Delicious small berries on easy to grow bushes. Needs a region or location that is not arid or extremely hot in summer or overly mild in winter (USDA zones 3–8). Hot summer areas in zones 7–8 likely won't work well for currants. Demand will be low to zero in many US locations, but would be a hit in others, so choose this crop with careful consideration as to local market. Great for value-added goods like jams. Make sure to grow the type in demand in your region. *Profit potential*: $20–50 per bush, maybe more in a strong market and with productive varieties.

Urban Market Farming Rating—4/5: Fairly space efficient per pound of fruit produced. If demand is high, this could be a solid urban farming crop due to heavy production and small space requirements. Bushes are somewhat compact and do not always require trellising.

Rural Market Farming Rating—5/5: If demand is there. These also fit in well in agroforest and orchard operations and between fruit trees. The plants are easy to cultivate but require lots of labor to harvest in mass.

Home Recommendations: 3–6 bushes should supply most households. Black currants may be the most productive of all. They all make excellent, healthy jams. In the southern range of their cultivation, plant in partial shade or areas with ½ day sun and ½ day partial shade.

Identifying Quality Stock: Lots of roots and at least an 8–12" top portion. Small plants as well as larger ones work fine and grow rapidly.

19

Figs

The fruit of heaven mentioned glowingly in every major world religious scripture from the Vedas to the Bible, figs are the tree to get enlightened under.[1] They're also an excellent small fruit crop to add to the market farm, but require some planning and special considerations, mostly around their lack of hardiness to frigid temperatures.

Fig trees are extremely well suited to organic agriculture. I'll often refer to them herein as "fig bushes" when grown in cold winter climates, which drastically limits their size. Figs are somewhat cold-hardy subtropical plants, but in most regions of the USA and Canada you *cannot grow figs outside in the ground without special winter protection*, which I will describe in detail later.

The delicious fruit is soft and perishable. It has strong demand in many areas, but may require some special marketing strategies. There are many types of figs out there and a number of different species. We're talking about *Ficus carica*, specifically the "common" fig. Common figs do not require any pollination to set fruit. There is something called a *caprifig*, and this is basically a male, pollen-producing fig tree that makes little to no edible fruit, but is required to pollinate most figs that are not common figs.

Mediterranean climates and similar regions including the warmer, drier parts of California and the Southwest grow the best figs by far. However, wetter, colder climates like most of the Eastern USA to the

PNW, and also cooler regions of Europe can grow nice marketable figs also. Cultivar selection is crucial, as some will thrive or at least produce fruit in a given region, and others will completely fail. This is especially true in the marginal fig growing, cold winter regions of zones 5–7.

Cultivation

Planting

Plant figs in spring only, utilizing either potted or bare-root dormant plants. Plant carefully as figs are prone to transplant shock, especially when leafed out. Fall planting can work in warmer regions, but the risk of winter dieback is there still, and they should be covered. (If planting in CA, see gophers in Pests and Diseases.) Dig large, wide holes, deep enough to accommodate all the roots, and 24" (60 cm) wide. Water the fig in deeply and mulch heavily.

Irrigation

Figs are very resistant to drought once established and do not require irrigation except in dry climates or in drought conditions. Irrigation is not needed in the Eastern USA once figs are established. High tunnel figs require regular irrigation the first few years and in hot, drought conditions. Irrigation during initial fruit setting can increase yields but too much will water down the flavor, oversize the fruit and can lead to fruit splitting and can ruin the crop. Overhead irrigation should not be used as it can encourage fruit rot. Stop completely or reduce by 50–75% all irrigation as soon as fruit is full sized and begins ripening. Utilize orchard tubing with 2–3 emitters per plant. Exact watering needs per season will depend on your region and soil type.

For market farming purposes we will split growing figs into two categories. Cold Region and Warm Region.

Cold region (USDA zones 5–7)

Is there such a thing as a "cold hardy" fig? There are *some* fig cultivars out there that have trunks and limbs with vascular cells that are substantially more tolerant to cold temperatures than others. Figs from tropical and sub-tropical areas such as India typically die back to the ground at temperatures of about 20–25°F (–7 to –4°C). Fig cultivars originating from mountainous and colder regions of Europe have more cold tolerance and resist freeze damage or trunk dieback until temperatures reach 20°F or below. Below 15°F (–10°C) most any unprotected figs will show damage or severe dieback, often to the ground. One important factor that increases a "cold hardy" fig's ability to withstand cold temperatures is how well the wood lignifies before cold weather arrives. Figs that are still green and not lignified will suffer much dieback when a freeze event occurs.

The root systems of hardy figs, if mulched heavily, nearly always survive winter in zones 5–7 even when the tops die back completely. If the root system and even just a few inches of trunk survive winter, it will vigorously regrow new branches in spring. This prevents fruit set on most cultivars, because most cultivars fruit only on 1 or 2-year-old wood that would have overwintered at least one season.

However, "Mount Etna" type fig cultivars from Italy are very cold hardy. This includes 'Chicago Hardy', 'Italian Letizia', 'Malta Black', and likely a number of others. 'Chicago Hardy' has the ability to fruit on new growth ("primocane fruiting" if you will) and does not exclusively fruit on branches growing from 1 or 2-year-old wood. Other Mt. Etna types can do this too. These figs can be grown in colder climates, even with little protection besides mulch. Yields are much diminished when figs die back to the ground and only bear on new, 1st season shoots, but figs are at least obtained, often in regions where typically they could not be. This will not be conducive to marketing, but to home production only.

Growing hardy figs in high tunnels in cold regions

In zones 5–7 with adequately long, sunny growing seasons, cold hardy figs can do amazingly well in high tunnels. Due to their origination in dry, sunny, hot, sub-tropical climates, figs are very well adapted to high tunnel culture.[2]

High tunnel figs planted 5' apart in our high tunnel. Note small square cut in landscape fabric and drip irrigation tubing with spikes.

For market farming purposes, it's unrealistic and unprofitable to attempt to cultivate figs *outside in zones* 5–7. The yields will be too unpredictable and generally very low. You'll need more predictable results and steady yields in order to market. This means a high tunnel is necessary.

One challenge with high tunnel fig culture is that some fig cultivars have a tendency to split, especially if over irrigated. The fruit splits open and bursts, thus ruining it. Another issue is whether the little hole on the bottom of each fig, which is called an "eye," is open or closed. Closed or tight eyes will exclude SWD, ants, and moisture better than others, and this prevents premature spoilage or souring of the fruit, as well as insect infestation.

Cultivar selection is therefore crucial and needs to be focused not only on cold hardiness and flavor, but selecting figs that have small or closed eyes and that don't tend to split. In our trials we've found 'Chicago Hardy' to be hardy, productive and reliable, without splitting issues and with closed/small eyes. 'Chicago Hardy' is always the best one in terms of production and quality.

The following cold-hardy figs are worth *trialing* in high tunnels in zones 5–7: 'White Marseilles', 'White Triana', 'Adriatic', 'Peter's Honey Fig', 'Paradiso', 'San Pietro', 'Malta Black' (very promising), 'Negretta', 'Dalmatie', and 'Black Triana'. Most of these figs originated in colder regions of Greece, Italy, and Sicily, often sourced from mountainous regions.

We've trialed many cultivars indoors and have found some *unsuitable* for zones 5–7 high tunnel culture. These include: most LSU cultivars,

'Ronde de Bordeaux', 'Violette de Bordeaux', 'Green Ischia' (may work in long-season zone 7 regions), 'Hunt', and 'Celeste'. This is due to numerous factors, including splitting, low yields, lower quality under indoor growing, very late ripening, etc.

Other, very popular figs, such as 'Black Mission', 'Calimyrna', or any of the figs grown commercially in California or for sale in the Western states will generally *not survive or perform* in zones 5–7 or in the Eastern USA. Mostly this is due to lack of cold hardiness, but some cultivars such as 'Calimyrna' require the pollination services of tiny wasps that don't live in the Eastern USA.

In general, the warmer (zone 6–7) urban areas of New England are better for figs than in zones 5–6 further south, such as Kentucky, Ohio, or Indiana. This is due to New England's less humid summers, less extreme and dramatic winter temperature fluctuations, and large urban areas providing protected micro climates and windbreaks.

Breba crops

A breba crop is the first, early crop of figs that begins to emerge in late winter to early spring on 1 year or older wood. The breba crop usually differs in quality from the late or main crop of figs. It is often larger in size, with figs sometimes twice as big as the main crop. The quality can be similar, better, or worse than the main crop—it varies.

In warm-climate areas such as California or Spain, breba crops are consistent and expected. In cold regions when growing cold-hardy figs, the breba crop is usually destroyed by winter freezes or late spring frosts. So do not expect to ever get a breba crop in zones 5 or 6, and rarely in zone 7, with the exception of some warmer micro climates such as in New York City. It's very unlikely that overwintered high tunnel figs in zones 5–7 can produce consistent breba crops.

Understanding high tunnel fig growing

In my experience, a *double-walled, well-sealed, and inflated* high tunnel creates a micro climate about 4–5°F (~2–3°C) warmer in winter than outside temps. In a high tunnel that is *not* well sealed, expect about a 1–2°F

(0.5–1°C) difference. Yet, protection from intense winter ice storms and desiccating, freezing winds also affords protection in winter far beyond a few degrees of warmth. I have recorded temps of around 10–12°F (–12 to –11°C) in the high tunnel yet the figs were completely unharmed. Perhaps the cold was too brief to cause damage.

The protected high tunnel environment is also *drier and less humid* than outside the tunnel, and keeps fruit-damaging rains off any ripening figs. Pest pressure is usually reduced, especially birds. Figs thrive in the hotter and drier high tunnel environment. Drip irrigation via orchard tubing and emitters (3–4 per fig) is all that is needed to keep them growing vigorously. Do not overwater but keep constantly moist. I have our irrigation set to drip every 3 days for 10 minutes, and turn it off completely during the peak of the fig ripening season. Too high a soil moisture can make the figs split.

Figs grown in high tunnels will break dormancy earlier than figs outside, because of the warmer indoor temps. Expect bud break 2–4 weeks earlier in a high tunnel. This is advantageous because subsequently the indoor trees crop 2–4 weeks earlier than outside figs. In KY 'Chicago Hardy' starts to crop around August 10th–20th inside the tunnel, and usually around early/mid-September outside.

The disadvantage to earlier budding is that a late spring freeze event could damage the emerging shoots if it drops to 32°F (0°C) or below *inside* the tunnel. So, be prepared to temporarily and quickly heat the tunnel during a cold snap. You can install a small wood stove and metal chimney in the back of the tunnel for as little as a few hundred dollars. This works well and is simple to install. It works best if you also have electric greenhouse fans to disperse the warm air. Another option is propane tanks outfitted with top heaters from the hardware store. This works well in a pinch and you can heat a 50–75' (15–23 m) tunnel overnight to above 32°F for about $10, or half a tank of propane. This gets pricey, so be prepared with a wood stove or electric greenhouse heaters. One grower I know uses special red-light bulbs that warm their very small, insulated tunnel. Remember, you're only heating the tunnel very occasionally during extreme winter cold snaps or late freezes. You're not heating it all winter! However, having a wood stove or heaters installed can also help pull the figs through unscathed if winter temps get near zero.

You could also install, at the time of building the high tunnel, a geothermic passive heating system[3] or build an awesome underground walipini.[4] Whatever the case, you have to be able to seal it up *very tightly* to avoid drafts. Having it double-walled with an inflation fan that inflates it like a balloon (a common practice nowadays to provide dead-air space and thus insulation value) helps also. This goes a long way to keeping temps moderated on cold spring nights.

In sunny winter regions like ours, the tunnel needs to be opened on cold yet sunny days, if possible. We close it only when temps are going below about 25°F. You want it to stay cold, but be moderated. On a sunny, cold winter day a sealed-up tunnel can quickly reach 100°F. This warmup might cause the figs to break dormancy earlier, which is very detrimental, and could lead to severe damage. So, you will have to manage the temps in the high tunnel during winter by opening the door on sunny days, and closing it up tight on nights predicted to go below 25°F. Automated, electric systems for this exist. Forgetting to do this once or twice won't cause any issues.

Despite the challenges, I've found high tunnel figs very easy to manage in zone 6 and branch/trunk survival rates are very high, usually 100% on hardy cultivars. Through simple and minimal winter management (mainly just keeping it open on sunny winter days and sealed up in the depths of winter cold) the figs bud out appropriately in early spring. Our tunnels are also painted with *white greenhouse shade paint*, which is very important for keeping temps from going extreme in winter and especially in summer. We have found white shade paint to be superior to using black plastic shade cloth netting; it's easier to deal with, more environmentally friendly, and cheaper. See the Resources section.

Figs usually require irrigation when grown in a tunnel. I have four spike-style emitters in the ground on each fig, about 4–6 inches (10–15 cm) from the trunk. I have these attached to ¾" orchard tubing that is connected to the solenoid system used to control the irrigation. I have the controller set to water the figs three times a week for about 10 minutes, which is not more than a gallon or so of water a week (our clay soil holds substantial natural ground moisture also). In winter our irrigation system is turned off and drained, at which point I bring in a garden hose and hand water on warm days, but only about 1–3 times per winter, as the clay soil in KY usually stays moist all winter (your region may differ). Don't allow the soil to dry out in the tunnel during winter or the figs could get damaged or die. In-ground moisture might suffice but keep an eye on it, and hand water with a hose accordingly.

Figs respond to a high tunnel environment differently than outside conditions, and this can affect the flavor, size, color, and yields. Purple figs tend to be lighter or greener in color indoors as opposed to out. Some figs that perform well outdoors in the South have been poor performers in a tunnel environment farther north. In our trials, high tunnel fig fruits get 2–4× as big as the same fruit on cultivars grown outdoors. Flavor is milder indoors but still good to excellent.

'Chicago Hardy' at its peak putting out heavy yields of plump, delicious purple fruit.

For most cold region growers, 'Chicago Hardy' is simply going to be the best cultivar. Perhaps test a few other promising cultivars that exhibit known heavy production, specifically those with early, consistent ripening and that have strong cold hardiness. Our second-best indoor performer has been 'Olympian', which produces heavy yields of medium-large figs of excellent quality and appearance, yet ripening a few weeks later than 'Chicago Hardy'. However, it has a large open eye which gets infested with fruit flies, so it has been replaced.

'Chicago Hardy' keeps ripening fruit in abundance until the tunnel freezes, which occurs in our area sometime in late October or November. Heating the tunnel during the first few cold snaps in autumn could lengthen the production season by a few weeks at least. This could also delay the lignification of the fig wood, so be careful. Likewise, a heater could warm the tunnel slightly if a severe freeze event is going to occur and the trees are not *yet* lignified.

Marketable yields occur in year 2–3 in the high tunnel, depending on growth and establishment. By year 3, I was getting 20–25 lbs. (9–11 kg) per 'Chicago Hardy' fig, planted 5' apart.

Hardy fig training systems

A hobby fig grower in zone 6, West Virginia, Dave Moore, utilizes a fascinating espalier fig growing system similar to that used commercially in Japan. Here is how Dave describes his process:

> Last year (2018) I planted a row of figs (a Mt. Etna type) in my unheated high tunnel, 10' apart. Pruned them each to 2 upright shoots. When those shoots got about 3' tall, I bent them over [horizontally], and secured them in opposite directions, utilizing a piece of pipe 6 inches off the ground. These pipes run the length of the row, about 80' (24 m).
>
> So, the fig plant's trunk comes up 6 inches vertically, and then each of its 2 vertical branches are tied down to the pipes and run in opposite directions, approximately 6 inches off the ground. Soon after this, most of the buds along the vertical branches started to grow upright, all along that flat top part of the horizontal branches. A few figs formed the first season.
>
> Late fall to early winter, before it got really cold, I cut those upright shoots off at two buds. So, by November, the whole row was just over 6 inches high, along the whole length of the pipe. At that point (the figs are now dormant and leafless, in order to protect them) I covered the row with large pieces of cardboard,

and reemay row cover frost blankets. Maybe about 3' wide along the whole row. Fairly easy to do, really. I put a few moth balls and rodent bait underneath for good measure.

It got close to 0°F inside the high tunnel this winter. In spring when I uncovered it, there was no winter damage to the plants. By summer, the shoots came up along the whole length of the horizontal trunk and have grown 6–8'. Many, many figs!! A real success. And a lot easier than wrapping individual plants.

I topped most of the vertical shoots at 5–6' by mid-summer to encourage ripening of the remaining figs, and that made a big difference. On many of the ones I topped, they ripened every fig on that shoot, no little green ones left.

Thanks, Dave! To elaborate on Dave's modified Japanese espalier system, I'd like to break down and highlight the most important aspects:

1. Once planted or established, figs are encouraged or thinned to grow 2 upright shoots very low on the trunk. Once they are about 2–3' tall,

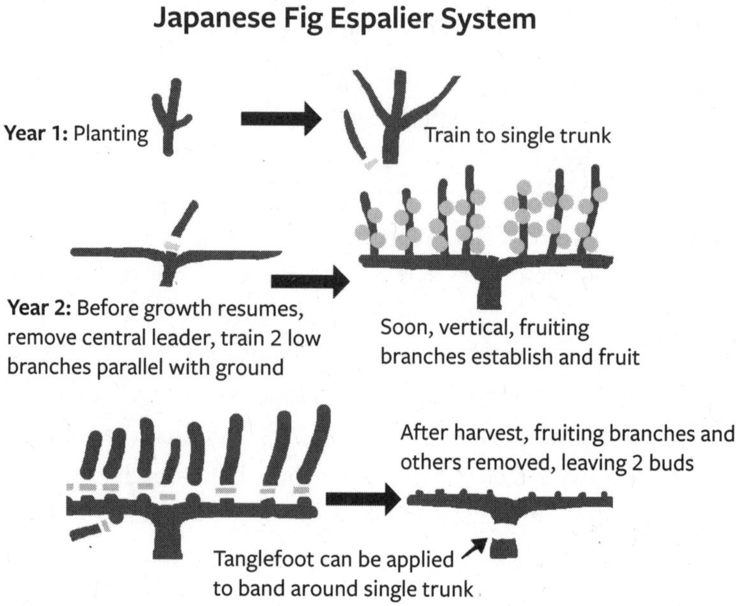

Japanese Fig Espalier System

each flexible shoot is tied horizontally to a horizontally oriented pipe/pole running parallel to the ground (presumably secured to the ground). These are the main branches. So, the fig was trained from a somewhat U shape to a T shape, with the bottom part of the T being the trunk, the top being the two branches going in opposite, horizontal directions. The bottom portion of the trunk can (and often should) be wrapped in a duct tape band and coated with Tanglefoot® for ant control. This easy ant control is a major advantage to the single trunk system. We successfully held branches down to a horizontal position with cement blocks instead of pipes.

2. The fig then grows many buds along those two main horizontal branches and they sprout and begin growing vertically, creating straight upright shoots. Those shoots will bear the figs. Thin these out to about 8 inches apart across the branch or they will shade each other out, diminishing yield and quality.
3. The vertical, fruiting shoots are headed back (stimulative pruning) to about 5–6' tall, to discourage further vegetative, vertical growth from hitting the ceiling of the high tunnel. This also helps the remaining figs to ripen.

4. After harvest and when the fig has gone dormant/leafless from cold weather, the vertical fruiting shoots are lopped off near the base, leaving 2 buds only. Make sure the 2 buds are left! These become next years' fruiting shoots. Cuts should be slightly sloped to drain moisture.
5. Before severe cold hits, the main branches and trunk are protected by blanketing them with large pieces of flat cardboard that is weighed down to the ground, which captures and holds ground heat. The cardboard is then covered in agricultural row cover (frost blankets), further insulating the figs. Use the most heavy-duty frost blanket available, and fresh, clean, thick sheets of cardboard. You may want to bait underneath the cover with rodent poison or traps if necessary. Wait as long as you can before covering, to lessen rodent problems. We've noted a 5–10°F warmer microclimate under this simple protection!

This training method is productive of fruit (and cuttings), is space efficient, and is easy and orderly to maintain. It also greatly simplifies protecting the figs for the winter. Definitely worth looking into this system if you're considering high tunnel fig production. We are now training all of our figs to this system. Can also be adapted to outdoor growing.

Spacing

High tunnel figs should be about 6–10' (1.8–3 m) apart, with about 6–8' between any rows. I have planted them as close as 4' (1.2 m) apart, but this is a little close. Pruned to bush size, or espaliered as described earlier, works well in the tunnel. If your tunnel is very tall and you want bigger bushes you might space them 8–10' (2.4–3 m) apart. When utilizing the Japanese method, figs are usually planted 8–10' apart. We are experimenting with having 3–4 main horizontal branches in an X type formation, as opposed to only 2 in a T shape.

Outdoor figs in cold winter climates should be about 6–9' (1.8–2.7 m) apart as they will remain as multi-stemmed bushes. *Outdoors in warm, mild winter climates where figs thrive, they should be planted 18–20' (5.5–6m) apart.*

High tunnel weed control, maintenance, and fertilizing

In the high tunnel environment we utilize landscape fabric, so weed control is nearly 100%. You may choose to do it differently. Weeds and grass can quickly get out of control in a high tunnel, so choose what will work best for you.

We fertilize each high tunnel fig with about a ½ pound (~230 g) of chicken manure (4-5-4) in early spring, and that's about it apart from a few kelp extract sprays mixed with neem, for micronutrients and as an anti-fungal foliage spray. Too much nitrogen and the growth will be ballistic with enormous leaves, 20-foot-long branches and minimal fruit production. *Be careful fertilizing figs!* Yellowing or pale green leaves usually indicates a need for nitrogen. If deficient, they respond quickly to small doses of liquid feed high in nitrogen.

When starting out new fig plants we feed them a bit heavier on nitrogen, so they rapidly grow a thick diameter trunk the first season or two. The aim is not a few fig fruits the first season, just strong vegetative growth and establishment. Larger fig bushes with thicker trunks survive winter better than little ones do.

In subsequent years, feed a moderate dose of nitrogen in spring. Don't overdo it! This is similar to growing tomatoes, which are sensitive to excess nitrogen as well. Figs respond positively to the addition of agricultural lime around the drip line and do not like acidic conditions.

Assure there is good airflow, otherwise the bushes will grow very floppy branches that tend to fall over. Utilize roll-up sides, or electric greenhouse fans along with louvres (metal slatted greenhouse windows) for airflow. If the branches are floppy they may need to be staked to bamboo poles or perhaps tied to the ceiling purlins. Poor airflow can also lead to fig rust (*Cerotelium fici*), a fungal infection that causes partial or total defoliation. Neem oil sprays help deter fig rust, as does providing good airflow, including removing excess branches and leaves.

Also make sure the figs are planted at least 2–4' from the side walls and end walls, so good airflow around the bushes in ensured. Don't let the leaves or branches grow so close that they are pressing into the plastic sidewalls or ceiling. Cut that growth off or tie it down more horizontally.

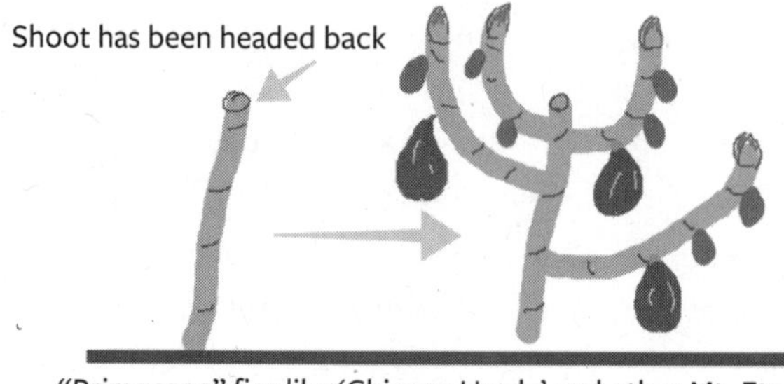

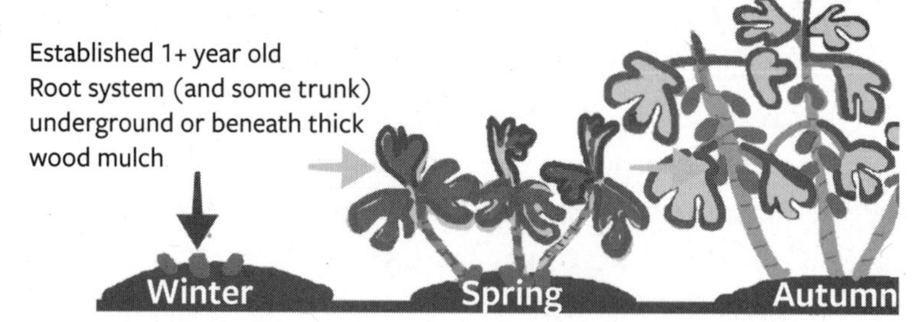

Pruning the tops of the branches in late summer directs energy into the remaining figs and helps them to ripen. Keep the tunnel very hygienic and clean of debris, including fallen leaves and, especially, rotting fruit.

Growing figs outdoors in cold winter regions (zones 5–7)

In cold winter regions, you will get one cycle of ripening figs from about August–mid October, making a picking window of about 3–6 weeks depending on cultivars and seasonal climate. In KY we pick for about 4–6 weeks outside, depending on season. Remember that most figs only fruit off of 1 year old or older wood. That means some amount of the branches or trunk has to survive the winter. Even "primocane" figs like 'Chicago Hardy' fruit much heavier if some amount of top growth overwinters.

To help assure this happens, all figs grown in zones 5–7 should be planted in a sunny, protected location such as near buildings (south side is best), or a wind-breaking privacy fence, etc. Never in an exposed, open location or frost pocket.

Some growers insulate fig bushes in autumn by creating chicken wire or other wire baskets around the entire bush and stuffing it full of insulating material (straw, leaves, etc.) A tarp tied over the top keeps it mostly dry. I did this for years and found it labor intensive and marginally effective. Now I do the traditional old-world technique of laying down the flexible branches and burying them within a small shallow trench, covered thickly with soil and wood mulch. Keeping them weighed down and buried is done using bricks or blocks. It's much less work, fairly easy to do and is very effective at overwintering the figs. We prefer this method. You can also chop the roots on one side and push the entire fig bush down into a dug-out trench, and cover with plywood and mulch (kind of like a fig tomb). The fig can be resurrected come spring by digging up and propping back up.

Traditionally many colder region growers grow figs in large containers and pots that are moved indoors for the winter via dolly. This is probably the most reasonable and practical method in cold winter areas.

Warm region (USDA zones 8–10)

Warm regions can be divided into two basic types: dry and humid. Warm, yet dry regions are generally Mediterranean or desert climates. Both are similar, except Mediterranean climates (such as California, the Southwest USA, or much of Italy) have a marked seasonal rainy period in winter, and receive more precipitation than deserts. These regions are perfect for growing figs, which do better in drier climates and produce better quality and drastically higher marketable fruit yields than in humid, rainy climates. Figs in this region will produce 100 times the figs as in the Eastern US. This is why the US fig industry is centered in Central California.

When growing figs in this region, irrigation will be needed. Also, choose premium cultivars, including ones that might require some caprifigs to be planted. The pollinating wasps should already be established in your region. If not, grow "common" figs.

Warm, humid fig growing regions receive lots of rain which complicates fig growing. These include the southern USA, Florida, and the Gulf of Mexico regions. See the end of this chapter for cultivar recommendations in this area. Cultivar selection is crucial.

It might be conducive to keep the figs cut back to about 8–10' even in warm climates, especially if you are harvesting for fresh market sales. This facilitates easier harvesting. Lots of heading cuts on the ends of branches in winter or early spring will stimulate fruiting branches to grow. Remove up to 1–2' per branch to stimulate fruiting branches and keep the trees in bounds. In warm climates, up to 3 or 4 fruit cycles on some cultivars is possible, with ripe figs possible almost non-stop from May/June–November.

Figs are marginal in wet, tropical climates and require very careful cultivar selection and are not described in this book.

In *zones 8–9* the only winter protection needed is on *newly planted, small figs*. A plastic barrel or old garbage can, etc., placed over the fig and secured via weight on top is all that is needed to hold in-ground heat and block winter winds from killing the young fig to the ground most seasons. After growing for a couple seasons this is no longer needed.

Harvesting

A word of caution

Fig latex is phytotoxic and can burn your skin! Especially the more potent latex from the broken or cut stems. Wear gloves and long sleeves if working with figs. On a positive note, fig latex is also very effective at destroying persistent skin warts, when applied several times daily for a few weeks to a month.

If you have a lot of figs to pick, wear gloves. Harvest figs only in dry conditions, as wet figs rapidly sour. If you must harvest wet fruit, then perhaps dry it rapidly by setting it in a single layer directly in front of electric fans set on high and placed near the fruit until dry. Pack or sell immediately after drying off. When harvesting, utilize wide, shallow containers for picking, so as not to mash the figs under their own weight. Stack no more than 2 layers high. Or, harvest directly into quarts and then put these in collapsible crates.

Figs should be harvested individually when fully colored and starting to droop on the branch, and should give slightly to a light squeeze. When picked at this stage, the fig usually drips a bit of latex from the detached stem. Figs picked for marketing *should not be showing any cracks.* If they are so soft they are showing cracks they will be too perishable to make it to market. They'd be perfect for home use or processing, perhaps a quick sale to a restaurant, but will not last at the market and will spoil rapidly. Experience will teach you just when they are ready. Picked firm ripe and allowed to sit for a day or two at room temperature, or in a fridge, renders them soft and delicious, something to inform your customers of. Figs picked *too* underripe will be inedible and will not ripen properly.

Here is a neat *Old World technique* for maximizing your fig harvests, that actually works. About 10–14 days before the first expected frost, take a bottle of olive or vegetable oil, dab a drop on your finger, and "paint" the eye of each large, green, unripe fig. For reasons unknown this causes the figs to rapidly ripen, within about 5–7 days. Granted, it sometimes blotches the end of the fig, discoloring it, and the flavor is slightly less than optimal, yet there are now substantially more fresh figs to be used, sold, processed, etc. *Viva la fig!*

Marketing

Fig demand varies across the US, but in many regions is strong. California, the Pacific Northwest, Southwest, as well as much of Texas, Florida, Mississippi, and Louisiana are all areas with good fig demand. New England (where many Italian and Greek immigrants and their descendants enjoy growing cold hardy figs) will also have good demand. The Midwest and upper South will often be unfamiliar with fresh figs, except areas in the Deep South where figs are more commonly grown (south MS, FL, LA). It's popular there to acquire figs for canning purposes.

Fresh figs need to be packaged nicely in pint containers or quarts. Fruits can be stacked only about 2–3 high. If needed, wipe them clean of debris or latex with a dry cloth before packing. Do not wash the figs with water.

Restaurants, specifically high end as well as Italian/Mediterranean restaurants, could be very interested in fresh figs. *As a rule, most markets prefer purple or dark figs*, but if the quality of your green or yellow figs is outstanding, you might convince them otherwise. Take some colorful photos and information to local upscale restaurants a few months before the figs ripen and try to make a deal. Figs are a rare gourmet item most places and should be priced accordingly, probably by pint or quart would be best. Per pint, $5–6 dollars and per quart $6–8 would be good prices. Maybe $4 per pint if a restaurant was buying a lot. Check current online and grocery store prices for fresh figs before deciding.

Direct marketing via the farmers market should be easy if any demand is there. People get excited about fresh figs. Sell via quarts or pints, priced as high as you can sell, probably $5–6 per quart, maybe more in an upscale area.

Value added goods

Figs lend themselves to an array of delicious value-added delights: fig bread, fig cookies, fig jam, fig butter, fig vinegar, fig paste, dried figs, fig cakes, fig smoothies, fig ice cream, fig syrup, and more. Wow! They're great as a topping on pizza. You could try making these products and marketing them, or give out recipes to your customers for inspiration and increased sales. Virtually anything made with figs is delicious and healthier. Figs are high in natural sugars, essential minerals, including iron, and fiber. They are known to be a natural encourager of bowel movements, gentle enough for babies, which is something needed in this time of depleted diets and low fiber intake. Thus, they are often an ingredient in natural herbal laxative products for that reason. They also are made into wine.

Fig pests and diseases

Diseases

Fig Rust: This fungal disease appears whenever humidity and temperatures are high, especially on susceptible cultivars. It creates orange, rust-colored patches on the leaves that eventually turn brown and necrotic and the leaves drop. In severe cases partial or near total defoliation can occur, although they usually push out more leaves after that. In humid regions, focus on *rust-resistant* cultivars. Often these are the traditionally popular cultivars: 'Celeste', 'Brown Turkey', 'Magnolia', and also ones developed by Louisiana State University (LSU cultivars). Neem oil applied as a foliar spray seems to help prevent rust in our tunnel. High tunnel figs are more susceptible, especially if air flow is poor.

Fig Mosaic Virus: This is a minor issue but is important to understand. Most available figs, especially nearly all of those from the West Coast, are infected with a virus that causes a few physiological complications, including distorted spring leaf growth and a lowering of fruit production. However, according to research, this only really affects the figs when the plants are small, because the virus is not rapidly mobile within the plant tissue. So, as the fig grows rapidly in the early summer it effectively "outgrows" the virus and appears quite normal and healthy (the virus catches up eventually). Mosaic symptoms will be obvious when growing infected figs from cuttings. The small starts may display misshaped, deformed leaves and leaf mottling, but once grown out appear fine. It is highly contagious, being transferred from infected plant to non-infected plant via aphids. Just realize most figs have the virus already, but try to acquire virus-free stock if available. We destroy any figs showing signs of mosaic and have kept our operation free of visible symptoms, as we do not grow any figs obviously infected, regardless of cultivar merits.

Fruit Split: A physiological disorder brought about by high rainfall or excessive or irregular irrigation. Fruits simply burst or split open from excess hydration, and thus rapidly ferment and rot. This can be minimized by growing regionally adapted cultivars (especially in high rainfall areas) and those known to resist fruit splitting, especially in high tunnels where the problem can be worse on susceptible cultivars.

Souring: A physiological disorder brought about by high humidity or rain soaking into the fig through the "eye" (the hole on the bottom). Insects can also make this worse by crawling into the eye. Figs that produce fruit with small or closed, tight "eyes" are less prone to souring. Most Southern adapted figs are necessarily of this type. Some exude sugar through the eye and seal it over.

Animal and insect pests of figs

Gophers: These large rodents will decimate unprotected trees in gopher regions, such as California, through feeding on the roots. The only controls are trapping and/or killing the pests or, better yet, plant your figs in impenetrable underground chickenwire baskets.[5]

Raccoons/opossums: I annually have nighttime fig thieves that inevitably turn out to be dexterous opossums and raccoons. Live traps work well for these pests. Place near the figs and bait with any of the following: marshmallows, peanut butter and crackers, slices of watermelon or cantaloupe, or dry cat food. Or contact your local animal control.

Birds: Usually a minor issue, but in severe cases bird netting may be needed. Cultivars that ripen to green and yellow are less targeted by birds due to the lack of visual cues (dark/brown color).

Insects

Ants: A minor to important problem on figs, indoors or out. They're a major fig pest for us. First, keep the area clean of weeds and tall grass. Hygiene is important—keep the trees well picked and free of rotting or buggy fruit. Remove and discard any fruit covered in ants. For severe cases, keep all trunks or branches protected by Tanglefoot® (see Chapter 9: Maintenance, Protection, and Organic Pest Control). Ants can destroy or damage a lot of figs, sometimes most of them. Some growers put cinnamon powder all around the base of the trees, and this might help.

Yellow Jackets: Same recommendations as for ants, except Tanglefoot® is *not* effective. Keep bushes well picked and clean, perhaps utilize yellow jacket traps.

Fig Fruit Flies: Can become an issue if figs are left to overripen or rot on the bushes, so keep things very clean. SWD can probably be counted on to join the party, so the same rules apply. If hygiene is maintained and figs are kept picked, this should be a minor problem.

Fig Bud Mites: These tiny arachnids can become a real problem very fast if allowed to establish, especially in a protected high tunnel environment, where mites and parasites can become heavy infestations quickly. *Be very careful where you source fig trees and cuttings*, especially when dealing with private collectors online. Symptoms include a mottling of leaf color, spots on fruits, and a strong decline in vigor. Once an infestation develops, it can get so severe that many people just get rid of all their figs for some time, allow the mites to die off (or apply pesticides to the area) and then eventually restart.

Take my advice on this: it's best to avoid the temptation to become a cultivar addict, sourcing cuttings from all over the place, and just stick to a few solid cultivars, especially when growing in high tunnels or for market. It's not a bad idea to treat all purchased dormant cuttings or plants with a dish soap detergent scrub down, disinfectant (H_2O_2), and organic pesticide (pyretherin) just before planting.

Ambrosia Beetles: These tiny, wood-boring brown beetles infest dead wood of fig trees, and if numbers grow they can attack living tissue as well, killing or stunting the tree. Hygiene is crucial: remove all dead branches, stumps, and damaged wood. Never leave prunings around, burn or compost immediately. When pruning larger growth and making large diameter wounds, it may be beneficial to paint the wound with latex paint or pruning sealer to help keep beetles out.

Cultivars

Below are Eastern USA cultivars that perform well under colder, humid conditions. These are all "common" figs meaning they require no fig wasps for pollination and will set fruit without them. There are too many to list here, but some of the more reliable as well as promising ones are below.

Cold hardy figs (zones 5–7)

Brown Turkey: There are a number of strains of this fig, including 'Texas Everbearing', 'Vern's', 'English', and a number of others. In general, 'Brown Turkey' is a reliable and cold hardy cultivar that produces small to medium sized brownish figs with good flavor. It handles humid conditions well and grows very fast. Considered one of the most cold-hardy cultivars. Make sure you get a cold hardy strain and not the 'California Brown Turkey'.

Chicago Hardy: This Mt. Etna type fig was discovered growing in a protected micro climate in the Chicago suburbs area in the late 20th Century. Produces quality fruit on new growth, meaning it does not require 1 or 2-year-old growth to produce fruiting branches. The implications are that it can die back nearly to the ground every winter and you can still harvest low yields of fruit the following summer on the new growth. It is cold hardy to zone 5 without protection (except thick mulch around the roots). Figs are dark purple, very delicious, rich and sweet with a tangy strawberry flavor and medium-small size. Overall, the best choice for zone 5–7 for both high tunnel, backyard, and market growing. Also, one of the best in the Deep South and handles hot, humid conditions very well.

Olympian: This newer cultivar was discovered fruiting prolifically in cool Olympia, WA. Apparently in some climates this fig produces purple figs, in hot summer areas like KY it produces greenish figs with a purple blush. The figs are plump, sometimes large, with good mild flavor and firm texture. Skin is fragile. Handles cold winters well with protection. One of the best cool summer climate figs and does very well in the high tunnel. The large, open eye is a disadvantage in the South as it is susceptible to insects and souring.

Malta Black: Delicious round purple figs, heavy production. Cold hardy Mt. Etna type. Has strong potential in high tunnel culture.

Kadota: An excellent greenish yellow thick-skinned fig with flavorful red/pink flesh. Adaptable and handles cold winters well. A popular, ancient cultivar.

Italian Letizia (Letzia): Another Mt. Etna type fig with a lot of potential in cold winter areas. Fruits are very similar to 'Chicago Hardy'.

White Marseilles: An unusual squat, shiny yellow fig with good, juicy, pear-like flavor. Ripens early, around late July. Seems fairly cold hardy but is probably best in zone 7 and higher. Lignifies its wood rapidly in autumn.

Violette de Bordeaux (Negronne): High quality fig with superior flavor. Has unusual foliage. Considered cold hardy. Did not perform well in our high tunnel trial.

Rhonde de Bordeaux: Very similar to Violette de Bordeaux. Did not perform especially well in our high tunnel trial.

Florea: A medium-small, greenish/brown fig popular in the Northeast. Has superior cold hardiness and good–fair quality. Ripens somewhat early, around late August. Flavorful and quite productive for a long season. Performed very well in our high tunnel trial, producing heavy yields from early August-frost.

White Triana: The fruit has green/pale yellow skin and red flesh. Delicious berry flavor. Fruit ripens in two stages; the first around mid-July in zone 5 and 6. Second crop ripens in September. The name originates from Italy in the Central Mediterranean.

JH Adriatic: A dwarfish fig tree the produces spherical green fruit with red pulp. Exceptional flavor and quality. Nearly indistinguishable from 'Green Ischia'.

Peter's Honey Fig: A Sicilian fig that ripens to yellow. Best in warm-summer areas. Quality is excellent, with thick skin and super sweet "honey fig" pulp.

Paradiso (Genova): Paradiso/Genova fig originates from the Northern Mediterranean, in the city of Genoa. This plant yields good crops of large fruit. The breba crop is a fist-size fruit, long shaped with white/golden skin and pink flesh. It is very sweet and juicy. This plant bears two crops in August and September. Zone 6, to 5 with protection.

San Piero: The fruit is large and long shaped with a dark purple/red skin and red flesh. It is very delicious with a sweet and juicy flavor. Possibly produces very early crops.

Celeste (Sugar Fig, Tennessee Mountain): An old heirloom Southern fig, this fig performs best in zone 7 and higher, although we have gotten figs, unprotected, in zone 6. Quality is excellent, with medium-small, sugary sweet greenish to pink-brown, honey figs. Drops figs in dry weather and rain ruins the figs, but still worth growing! Exceptionally cold hardy.

Nero 600m: Stands for where it was collected: Nero, Rome, 600 meters high. This cold hardy fig produces plump blackish figs of excellent quality with pink flesh. Handles cool summers well.

Warm Climate Cultivars (zones 8–10)

Again, there are far too many to list here, but a few popular, common, and notable ones with good marketing potential (or current marketing demand) are listed below. If you are interested in growing figs in warm climates, there could be better choices than some of these, although many of the best tasting ones may not be the best for marketing purposes due to having very soft fruit or lack of demand, or low productivity. All of the following figs are 'common' figs except 'Calimyrna'.

Black Mission: A superior fig and one of the top commercial cultivars, common in California, where it originated from the Spanish missionaries. Fruit is rich, very sweet and dries exceptionally well. Figs are dark purple, almost black with bright red flesh. Top choice and an excellent, highly in demand market cultivar.

Calimyrna: Originally from Smyrna, this is a top commercial fig in California. Requires a special fig wasp for production, and careful cultivation techniques. Ripens to yellow/tan and has superb quality.
Conadria: Very adaptable cultivar ripens to green and has amber/pink flesh. Excellent rich flavor, one of the best. Dries exceptionally well. Does well in cool coastal areas as well as hot summer areas. Performs very well in hot, humid, deep south conditions as well. Excellent market cultivar.
Desert King: One of the best figs for cool PNW climate zones, and much of California due to ripening well in cool summers. Ripens to yellow-green. Rich, sweet flavor and excellent quality.
Osbourne Prolific: Again, one of the best figs for cool PNW climate zones, and much of California due to ripening well in cool summers. Figs ripen to purple/brown and flesh is amber. Quality is very good.
Tiger Stripe/Panache: Has striking white and green stripes on the skin. Ripens to green-yellow and white. Flesh is dark red. Delicious, rich strawberry flavor. Excellent market cultivar; stays firm when ripe, so handles packing well.

Hot, Humid Climate Cultivars (zones 8–10)

LSU Purple: Resistant to fig rust, cracking, and rot, so performs well in wet summer areas like the Deep South. Figs ripen to dark purple, are small and have a pale amber flesh. Quality is good, some say excellent. Zone 8 and higher.
LSU Tiger: Resistant to fig rust, cracking, and rot, so performs well in wet summer areas like the deep south. Figs ripen to reddish with red stripes, are medium-sized and have a deep red flesh with excellent berry flavor. Said to be good for cool summer areas also. Zone 8 and higher.
LSU Scott's Black: Resistant to fig rust, cracking and rot, so performs well in wet summer areas like the Deep South. Figs ripen to dark purple/black, are medium size and have a red flesh. Quality is excellent and better than many other LSU selections. Zone 8 and higher.
LSU Improved Celeste: Developed from 'Celeste', this fig is very tasty and productive and shows decent cold hardiness, though is probably best in zone 7 and higher. Produces early crops. Fruits are plump, thick

skinned, and flavorful. High quality. Several strains available, with varying quality.

Magnolia: Juicy, flavorful, reddish-purple/brown when ripe. The figs handle rain and humidity very well and are not prone to cracking. Heirloom cultivar used for canning, but also good fresh.

Green Ischia: Quality is superb especially for a somewhat cold hardy fig. The thicker skins on this fig resists rain and humidity extremely well, making this one *very well suited to the Southeast and even the deep south—thrives all the way to the Gulf.* Figs ripen to yellow-green, making bird predation less. Inside is a dark reddish pulp with intense strawberry flavor and sweetness. Very productive. Not to be overlooked.

Texas BA-1: Superior fig in hot, humid climates. Rare in the nursery trade, but sometimes available. Good quality fruit similar to Brown Turkey.

Smith: A superior fig for growing in hot, humid climates. Uncommon in the nursery trade but is sometimes available. Makes a roundish, black fig with red flesh. Exceptional, rich flavor. Heirloom from Becknel Nursery in Louisiana. Excellent marketing fig.

Italian Black and Native Black: Both superior dark purple/black figs for growing in hot, humid climates. Rare in the nursery trade, but sometimes available. Excellent marketing figs. From Becknel Nursery in Louisiana.

Sicilian Hardy: A hardy cultivar that handles hot, humid conditions very well. Produces round, golf ball sized yellow honey figs that are very sweet. Rare.

Also recommended for hot, humid regions: 'Miss Hall', 'Celeste', 'Chicago Hardy', 'Conadria', and most other LSU fig cultivars.

The following conversation is an impromptu chat I had with my friend Stan, a successful, small fruit market farmer and fig grower in Picayune, Mississippi. Stan is a very detail-oriented, diligent, and observant grower with a penchant for consistent high quality and exact details in his work. Picayune has a brutally hot summer with extreme humidity and occasional summer rain and storms—not traditionally considered good fig growing territory by any means. Winters are wet and cold but fairly mild, zone 8b. He has been growing and marketing his figs for about 20 years or so. He also markets Asian persimmons, muscadine grapes, satsumas and other produce. He tends to about 50 small, pruned fig trees, mostly in his large backyard, so this is definitely a micro-farming type endeavor, yet financially substantial for him, and has proven easy to manage and also lucrative. According to my estimates he is bringing in approximately $5000 a year on figs and fig products alone, on less than an acre, and as a small side gig he greatly enjoys doing. This number also leaves out the other fruits he produces and sells on this same acre or two. In 20 years of doing this, that would equal approximately $100k in revenue, just from fifty fig trees! At 70 years old he is still enthusiastically at his work, even in failing health.

Blake: Which figs are the very best performers in this area (near the Gulf of Mexico in hot, wet, zone 8 Picayune, MS)?
Stan: I've had good success with 'Green Ischia', 'Texas Everbearing', 'Texas BA-1', 'Smith', 'Conadria', 'LSU Purple', 'Miss Hall', 'Italian Black' and 'Native Black'. 'Chicago Hardy' also does especially well here. I'm trialing 'Papa John', 'Ce Bonn' and 'Olympian'. 'Papa John' does real well so far. 'Italian Black' and 'Native Black' were very old heirlooms saved and propagated by the Becknel brothers in Louisiana. 'Celeste' and 'Conadria' are the 2 earliest ripening figs that we have. Closely followed by 'Chicago Hardy', which produced 3 crops last year. A trait of 'Celeste' I don't admire is its tendency to shed figs if the weather turns dry. Other varieties don't do that. On others the figs are just smaller due to insufficient water. 'Celeste' has an unbeatable flavor but easily spoils in wet weather.
Blake: What has been good about these adapted cultivars?
Stan: They handle our wet climate here without souring and rotting. The fruit is good quality, abundant, and has been good for marketing.

Blake: How have you determined what cultivars work out here?
Stan: I've grown many different cultivars over the years. Generally, I trial each one for 4 years to test how it grows and how the fruit handles our conditions. It takes a while for figs to produce mature fruit. You can't tell how good it will be until they have been growing for a number of years. But if every year it rots on the bush or never tastes good, I remove that one and try another. So, I give them 4 years each, and don't spray anything on them. I've narrowed down the best ones for marketing to this list.
Blake: What makes these particular figs good for marketing?
Stan: Purple and black figs always are in higher demand than green or yellow ones. People prefer darker figs. *We've had to really educate people* about how good yellow and green figs like 'Green Ischia' are. Now, 'Green Ischia' is the favorite of many people. But in general, the darker ones are in greater demand. These figs stay good in our wet summer weather and harvests are heavy. The quality is very good with these figs and so they sell readily. I had one restaurant that always requested 'Conadria' figs; they're so big and easy to work with, and they're so sweet. During the summer, their quality is just amazing. I mostly sell to restaurants and high-end grocery stores.
Blake: How do you grow your figs?
Stan: It's best to plant them in the spring. I plant the trees on 20-foot centers. I use lots of mulch, usually old hay. Never use Bermuda grass hay as it can re-root into the ground and it's a huge pain to try to get it out of your orchard. I water the trees by hand when it gets really dry in the summer. In winter I prune them all to about 7–8' high so picking is easy and they stay within bounds. I pot some of the cuttings and sell about 800–1000 every year. I usually sell them for $3 each to a local farm store who retails them. Our local population in Picayune is low income and people won't pay more than $5–6 per tree. I upped the price a few dollars one year and sold about 80% less trees, at which point I went back to $3 per tree.
Blake: What do people do with figs out here?

Stan: A lot of it goes into making fig jams and preserves. That's why people like the darker figs; they make a nice dark jam. The light-colored figs also darken when they are cooked.

Blake: Do you have any problems in the orchard with nematodes? I hear they can really devastate figs out here in this sandy soil.

Stan: I don't really have major problems with nematodes. I had one tree mysteriously die last season, and I'm guessing it was nematodes. Armadillos can dig into the mulch and cause a lot of damage sometimes. I used to have a pit bull that would catch and snap armadillos in half like crackers. Dogs help with those.

Blake: Are birds an issue?

Stan: Not so much but they can sometimes peck at the fruit. The green and yellow cultivars are less predated by birds.

Blake: Is any winter protection needed out here in zone 8b?

Stan: Mature trees are fine. But, when I've planted out new trees in spring and winter is approaching, I take an old plastic trashcan or barrel and cut the bottom out, so it's hollow like a tube. I place this over the tree, kind of snugging it into the mulch layer. This holds in some heat and blocks harsh winter winds and they pull through winter no problem. I've found many times small figs die back to the roots if not protected like this. Then it is harder for them to establish big enough to not die back every succeeding winter. When planted in spring and protected the following winter they are usually big enough by the following winter that they do not require any further protection.

Thanks, Stan!

Summary

A timeless fruit for all ages, figs deserve much wider growing and utilization. The fruit is easy to grow, abundant, healthy, versatile, and delicious. The trees are highly adaptable and very disease and insect resistant. In tunnels they can be grown in cold climates successfully. In the South and Western US this crop is already popular, as it is in parts of New England.

In other areas it may need to be thoughtfully introduced but should catch on quickly. Excess fruit makes excellent value-added goods. *Profit potential*: This will vary drastically from one region to another. In Southern California, where figs grow large and are extremely productive, yields can average a mammoth 788 pounds per tree and profit estimates are $2599 in gross revenue.[6] With cold hardy figs in high tunnels, especially 'Chicago Hardy', we harvest about 20 pounds of fruit per 'Chicago Hardy' bush, worth about $80–160 depending on direct market pricing. According to the University of Kentucky, 10 pounds of hardy figs will need to be produced per plant and sold for (the very low price) of at least $3 per pound for growers to make a profit.[7] In mild winter areas such as the Deep South, where figs can be quite productive and large (but not nearly as productive as in CA), estimates are about $100–250 per fig tree in a good season, maybe more, direct marketing fresh figs. It's unlikely that growing outdoor figs in zones 5–7 could be profitable at all, due to the very late ripening of the fruit in these regions (late September–October) and thus very short picking window (2–6 weeks)before freezing weather forces the bushes dormant.

Urban Market Farming Rating—4/5: In appropriate warm climates they can be maintained outdoors as large bushes and yields are high. Growing them in tunnels in an urban setting will not be efficient on a dollar yield per square foot basis compared to more profitable crops such as tomatoes, greens, etc.

Rural Market Farming Rating—5/5: Easy to grow, and without space restrictions this could be an excellent niche crop to get into. Make sure quality is high and demand is there. Tunnel grown figs are profitable and versatile. Far northern regions will likely not have sufficiently long summers to make this crop very profitable, even in tunnels.

Home Recommendations: In zones 5–7 focus only on the most cold-hardy cultivars and plant in the sunniest, yet protected location possible or in a high tunnel. Next to a sunny, warm brick wall, garage, or large compost pile works well. Mt. Etna cold hardy types should be a focus. In zones 8–10 plant cultivars adapted to your region in a sunny location. Keep mulched and irrigated at all times in dry regions. Don't plant near septic or other pipes.

Identifying Quality Stock: Plants should be at least 1–2' tall. Bare-root or potted are both fine. Lots of orange-yellow roots should be present and the tops and stem should not be mushy or rotted. Seeing a tight, hard bud on the very top is ideal. Dormant plants handle transplanting better. Quality cuttings work fine too, and grow rapidly, fruiting in only 1–2 seasons from planting.

20

Tomatoes

Tomatoes are technically berries and are the seed-bearing fruit of the nightshade plant, *Solanum lycopersicum*. They can be a real workhorse and quite profitable for the ambitious and diligent grower. Also worth noting, bell peppers and eggplants, including Asian eggplants, have very similar requirements as tomatoes and are also good niche fruit crops. They are much less productive than tomatoes but require less care and maintenance, and can be profitable.

There are a few major considerations with growing tomatoes for profit.

First, you have to plant regionally adapted cultivars. In the South and Mid-Atlantic, it's important to have strong disease, moisture, and heat resistance. In the Deep South, only a few select heat-tolerant and disease-resistant cultivars should be attempted, and with February transplanting. Up North, you need vines that can handle cooler summers, shorter summers, and ripen fruit in less heat.

Second, do not over-emphasize heirlooms. These fad tomatoes look gorgeous in magazines, but *production, consistency, and resiliency* are often lacking for the market grower. You're doing this in large part to make a profit and keep the farm alive, not bragging rights and photo-ops. We must question, what is it that people *actually* like most about heirlooms? It's fundamentally the unusual colors and good flavor. If they tasted terrible, no one would care that the cultivar is old, or has purple zebra stripes, etc.

Currently there are a number of considerably "heirloom-like," modern, non-GMO, F1 hybrid, and also open-pollinated cultivars available today, often under different trade names by different companies ('heirloom marriage', Wild Boar Farms,[1] etc.) These merge greater vigor, reliability, enhanced disease resistance, and better production with substantially heightened fruit quality and a colorful, attractive look. We grew a number of these types and they sold like hot cakes at the farmers market. Everyone assumed they were heirlooms, and the exceptional high quality the customers got and the high price we received left everyone satisfied. These come in all colors and shapes, including brown/purple, yellow, striped, etc.

Third, know your market and plan accordingly. Does your local market accept only red tomatoes? Is your market saturated with heirlooms and lacks regular reds? Does your market eagerly buy funky, colorful, weird tomatoes? Are both in strong demand, and if so, is one of these in short supply? Are cherry tomatoes viable or a waste of space? How many tomatoes is enough? If you have extra, where can you rapidly move them in quantity so they don't spoil? These questions need careful consideration.

Cultivation

In humid, wet climates like the eastern USA, and especially in the Southeast, I would say it's better to over-plant tomatoes by maybe 10–15% of what you calculate is "enough." The reasons being that, sometimes, prolonged wet weather can ruin a lot of tomato fruits; vines sometimes spontaneously die off; raccoon (and other) theft occurs; sometimes cultivars have a bad season, or fail to perform, etc.

In general, it's most profitable to produce hybrid red slicer tomatoes. These remain by far the strongest sellers. The "Mountain" series of tomatoes is a great place to start ('Mountain Pride', 'Mountain True', 'Mountain Spring', etc.) These are reliable, good quality, resilient, and very productive. Flavor is decent to good. If it's big, red, and local, that's honestly enough for most people; it doesn't have to be a flavor party.

If you are set on growing non-red tomatoes, I would make them no more than 10–25% of your planting until you are sure they will be a sure-fire success. Some folks have made their farms famous through the funky tomato fad. Usually this is in areas well acclimated to growing tomatoes (read: drier climates) and with bountiful urban markets. Another niche product could be high quality cherry tomatoes. These come in all colors and sizes. Make sure the cultivar you choose will be large enough in size and very productive. Check with your local ag extension office to see what cultivars they recommend for *commercial* production (not home gardening).

'Sakura' cherries are very tasty and productive.

Starting plants

I have found it best not to start the seeds too early. About 5–8 weeks before your planting out date is good. If you are going for extra-early production (or are in the South) you will start them in December and transplant into a heated high tunnel in Feb-March (or outdoors in the South).

Starting tomatoes outdoors in greenhouses is risky in winter or early spring because they can easily get killed by cold. We start them indoors, planting seeds in 72-count cell trays in well-draining and fertilized potting soil in early March to plant out in early May. Vermiculite in the potting mix helps hold moisture. Starting them under LED grow lights put on timers (on from dawn to dusk or longer, up to 16 hours or so) works great. Position the lights just above the plants.

Keep fans on the plants all day blowing them gently around so they stay healthy and grow strong. These can be put on the same timers as the lights. Water every 3–4 days as needed via pouring an inch or two of water in the flats holding the cell trays (solid flats only, with no drainage holes).

Tomatoes prefer cooler temps at night if possible (60–65°F) (16–18°C). After sprouting and growing for about 1 month in the cells, transplant into larger cells or 3–4" cube pots. When transplanting, always remove lower leaves and bury the stem deeply under the soil, about 75% of the entire stem. The stem rapidly produces new roots for stronger establishment and more vigorous growth. If they get 'leggy' (stretched) on you, transplant very deep into their final location and they will recover.

Keep the plants lightly fed with a highly diluted (25% strength) balanced liquid vegan organic fertilizer, yet not overfed to where they grow rank and overly lush. Start some extras in case of problems, and sell any excess. This could be a viable side income at the market itself. Vigorous side shoots found in between the branches can be removed when 6–8" tall and planted to fill in gaps or to make an additional, later ripening planting.

Training

There are always new trellis systems being developed, so check into the newest tools and techniques. Typically, high tunnel tomatoes are planted closely (10–12" apart), heavily pruned by removing every sucker (next to the leaf bases) so only *one main stem* grows, which is trained up a string or wire hanging from a special spool attached to the rafters or purlins. When the plant reaches the top, the twine is let loose from the spool and the vine is lowered down towards the ground, where it can grow back up the string again. That way, indeterminate tomato vines can keep on producing fruit for as long as possible, 3–4 months or so. Removing all bottom leaves as the vine grows improves air flow and reduces disease pressure, and allows multi-cropping shade-tolerant veggies at the base of the tomatoes (lettuce, basil, beets, carrots, herbs, etc.).

Outdoors, the "Florida weave" system of trellising is ideal.[2] Factor in the costs of wooden tomato stakes, 1 stake in between every two vines (50–52 per 100 vines). Organic certification prohibits using treated stakes, so go for white oak, pine or similar, or metal T-posts. Make sure spacing is correct for your area (closer in dry climates) and prune very heavily for good airflow. Pruning requires considerable labor, about 2–3 minutes per plant per season.

Determinate or semi-determinate vines may be better in outdoor cultivation, otherwise the patch can turn into a mad jungle of enormous sprawling vines. Watch out for raccoons and have live traps on-hand. Consider using organic anti-fungal sprays in humid climates, keep pruned of lower leaves and bottom shoots, and always choose disease-resistant cultivars only. Green tomatoes are also sometimes marketable, and if vines start to die off or green tomatoes break off, you can still harvest and market those.

Some people are experimenting with autumn tomato cropping, aiming to provide these when local tomatoes are few in number and prices are higher. Plants are started in June or July and planted in tunnels in August, or further South, September. Plants will crop until it freezes in the tunnel. Heating is possible and likely necessary in some situations.

Strategies

The local Amish have developed a high-yielding tomato growing system. They focus on growing potted, semi-determinate, large fruited, modern commercial hybrid slicer tomatoes in high tunnels. High tunnels provide a drier, warmer climate that tomatoes thrive in. Second, they grow each tomato plant in 1 gallon plastic grow bags or reusable plastic pots using sterile potting soil. Fertilization is via drip fertigation and the plants are trained up vertical lines on spools, with careful and very hard pruning. Purchased bumblebees help with pollination. Yields are super high.

Their family labor being free, a 30' × 50' high tunnel managed in this way can bring in $2000+ per week for about 1–2 months, marketing the fruit at wholesale prices. They do this through local Amish-operated produce auctions, receiving on an average about $1–2 per pound, but sometimes as high as $3 or more. Their substantial profits would be even better if the tomatoes were direct-sold at retail pricing. According to one family, they make $2000+/week off roughly 500 plants grown in pots in one high tunnel, for at least about 4–6 weeks of the season, selling

Note the tomatoes in pots and with support wire ties to purlins. Plants in this tunnel are fertigated and also sprayed with organic anti-fungal products.

wholesale. For another 4–6 weeks they continue to sell the tomatoes but by then the market is saturated and prices drop. Their aim is to have ripe tomatoes in early March, when prices peak, which necessitates starting plants in December and growing in heated high tunnels, which they power with Amish lumberyard scrap wood or sometimes coal, also utilizing sophisticated heat distribution ceiling tubes. Solar panels keep their tunnels inflated via fans.

Indoor growing similar to this could be a great niche strategy to look into, and could work even better in an urban setting with bustling markets. Although the Amish growers are usually conventional (non-organic) this system could easily be adjusted for organic inputs in the potting soil, fertigation, and any sprays.

You can also grow in-ground inside high tunnels with great success, though you'll have to rotate crops annually, only growing tomatoes again after a few other crops have rotated through (ex: tomatoes to cucumbers to high value greens back to tomatoes, etc.). You might build two or more tunnels so you can rotate crops between them.

Pests and diseases

This is a large topic beyond the scope of this book, but a few notables are below:

Deer and raccoons: Deer will eat vines and fruit. Fencing is often necessary. Raccoons can also become seriously addicted tomato thieves. Live trapping and relocation may be necessary.

Tomato Hornworms (*Manduca quinquemaculata*): These enormous green caterpillars can defoliate an entire tomato in hours, and also eat green tomatoes. In many regions they readily get parasitized by a parasitic wasp, and once the wasp is established, your problems with these will be mostly over. So, handpick and relocate well away from the plants. If covered in white (wasp) eggs, do not disturb.

Army worm (*Spodoptera eridania*): This dark brown small caterpillar can become a serious pest in mid-late summer. You'll see ragged, shallow holes in the fruit and brown droppings. Organic pesticide sprays can help.

Tomato Blossom End Rot: Physiological disorder (*not a disease*) caused by calcium uptake deficiency,[3] remedied by providing consistent moisture to the plants, and mulch outside.

Marmorated and other stink bugs: Currently these might prove to be a major challenge. You may need to resort to neem oil and other organic insecticides (pyrethrum) to protect the fruit. Damaged fruit displays tiny, hard, circular green spots on the skin that never turn red. The fruit may still be marketable but this lowers the appeal and disqualifies them from grocery stores.

Most other issues revolve around *foliage diseases* and various *fungal blights*. Good trellising, adequate space, hard pruning for strong airflow, planting strongly resistant cultivars, providing some mulch, and high tunnel growing help with prevention. Also, there are many anti-fungal organic products on the current market that can be very effective. Contact organic pest control product companies for their newest recommendations. Many modern cultivars developed in Florida and other southern areas have especially strong built-in disease resistance. *It's worth looking into these!*

When harvesting, use snips. Don't tear them off! Maybe grape snips would be a good choice for tomatoes. Keep the green top, but cut the stem very short so it doesn't poke holes in the other tomatoes in the crates.

Irrigation

Tomatoes are fairly resistant to drought on healthy high organic matter soil. Once established they do not require irrigation except in dry summer climates or in severe drought conditions. Irrigation during initial fruit setting can increase yields but too much will water down the flavor and can lead to root diseases and fruit split. Overhead irrigation should not be used as it can encourage fungal diseases. Reduce irrigation as soon as fruit is full sized and begins ripening. Utilize one line of drip tape per row. Exact watering needs per season will depend on your region and soil type. Irrigation is required in high tunnels but not in field growing in the Eastern USA most seasons. However, it may be a good idea to have drip tape or irrigation supplies on hand in case a severe drought hits. Allowing the soil to get very dry and then very wet will cause splitting and cracking of fruits.

Marketing

Pretty straightforward. Direct market as much as possible at farmers markets. Local restaurants often get inundated early on, so consider that. $2–3 per pound has been the going rate for fresh tomatoes for some time now. In really hot markets, with excellent quality, you may get $4–5 per pound. Per 25' indoor bed when well grown around $700 in returns is possible. Cherry tomatoes should be sold in quarts and pints. Others by the pound. Do not stack the fruit, harvest one layer deep only in collapsible harvest crates. Wipe them clean with a soft rag before displaying at market.

Bruised and unmarketable fruit could be made into value-added goods. Also consider those fruits for harvesting seed, which is a profitable product in itself. As a very casual side item we sold over $900 in quality organic tomato seed online in 2021 from our garden.

Cultivars

Focus on cultivars that can produce nice marketable tomatoes in abundance under organic conditions. Many heirlooms are amazing but seriously lacking in the production, disease resistance, and reliability departments, making them oftentimes a poor choice for the serious market grower. Also, unexpected summer rains can ruin them easily, making them crack and rot and infecting the foliage with deadly blights. I speak from decades of experience growing tomatoes both in the field and in high tunnels. If you want to make money doing this efficiently, focus on growing high quality F1 hybrids, which is advice I disagreed with at first, and later wished I had listened to. Alternatively, Wild Boar Farms cultivars are as good or usually far better than most heirlooms while still being colorful and weird. They are also open-pollinated for seed saving. If you come across an heirloom red slicer cultivar that produces as reliably and heavily as hybrid slicers, perfect. As far as this goes, especially in challenging wet growing conditions, I have not.

Mountain Series: There are many cultivars in this series. In general, these workhorses are large, round, red slicer tomatoes with strong production, disease resistance, reliability in wet and cooler areas, with fair to good flavor. Their excellent yields of large, meaty, firm, high-value marketable fruit will make your market table look stacked. Consistently our most productive and survivable tomatoes, still producing good fruit by the time others have died out.

Wild Boar Series: If your market calls for lively colors, heirlooms, and funkiness, check these out! More reliable with better disease resistance than heirlooms but with equal or much better quality. I especially like 'Pink Berkeley Tie Dye'. It was a hit with our customers and performed very well.

Hybrid "heirloom lookalikes": These types of hybrids are a great choice for heirloom lookalikes that yield and survive much better than "real" heirlooms, yet look and taste just like the old-fashioned ones. Many of these superior hybrid cultivars are currently available from a number of seed companies. Trial a few and find which ones perform best.

Marnero (F1): Looks and tastes exactly like 'Cherokee Purple' but is a vigorous and productive hybrid from Johnny's Seeds.

Sungold: A tasty, productive, yellow cherry tomato.
Sakura: A productive and medium-large hybrid cherry tomato with excellent flavor.
San Marzano: An heirloom sauce tomato with reliable and heavy production of large tomatoes with excellent quality.
Solar Set: Hybrid, determinate, 70 days, bright red globe (9–10 oz./255–284 g), excellent heat and disease resistance for the South. Developed by the University of Florida
Sunmaster: Hybrid, determinate, 72 days, red globe (7–8 oz./198–227 g), strong disease and heat resistance for the South, and is said to produce fruit even at temperatures above 90°F.

Summary

Universally popular and high value fruit beyond all temporary fads. Always in strong demand. Fast and easy to produce. Cultivar selection is crucial—hybrids are a great choice. *Make sure to carefully choose determinate or indeterminate* depending on your grow system. Tunnel growing is a great option and can greatly increase yields and quality. Tomatoes are disease-prone in humid areas so choose disease-resistant cultivars. *Profit potential*: High, especially when well-grown in a high tunnel. Estimates of $15–25 per plant when well grown and direct marketed at high prices.

Urban Market Farming Rating—5/5: Popular and easy to grow. Disease and insect pressure should be lower in a city. Growing them in tunnels in an urban setting is highly efficient on a dollar yield per square foot basis. Find your open local niche and fill it.

Rural Market Farming Rating—5/5: Easy to grow. Make sure quality is high and what you produce is locally in demand. Choose productive, disease-resistant cultivars only. Tunnel grown tomatoes are profitable and yields can be big.

Home recommendations: Focus on the end goal (processing or fresh eating) and plant resilient cultivars that can handle your local climate, be it cool, hot, very hot, humid, dry, etc. Hot humid regions require early planting and cultivars with very strong disease resistance. Trellis and prune as explained earlier. Check with your local ag extension office for recommendations.

Identifying Quality Stock: Plants should be bright green, potted, and not bare-root. Plugs work well. Plants can be anywhere from 3–12" tall. If tall, bury most of the stem at planting. Small plants work fine and grow extremely fast. No flowers or tiny fruits should be present at planting or purchase. If they are, immediately remove them. You might also consider utilizing special grafted tomato plants,[4] which research has indicated produce higher yields.

21

Other Small Fruits Worthy of Consideration…Or Not

Aronia (*Aronia melanocarpa*)

The ripe, dark purple fruits of selected cultivars are the size of large peas and are nearly inedible raw but when cooked and sweetened are delicious. Blue Fruit Farm, whom we feature in the Interviews section, is successfully farming and marketing this fruit in Minnesota. There is little in the way of pests or diseases and they are very easy to grow. They start producing fruit in about 2–3 years from planting.

As an easy-to-grow, native, very productive and multi-use plant, it has a lot to offer. With some breeding they could likely be sweetened and the fruit made larger, too. US growers are already benefitting from cultivating aronia for the health food and supplement trade.[1] It's apparently an important cultivated crop in Poland. Some midwestern US growers have struggled to make this crop financially lucrative.

Future Market outlook: These have potential mostly in *northern, Mid-Atlantic, and Mid-Western prairie regions* but lots of customer education is needed to sell the fruit, or a wholesale buyer of dried berries is needed. Value-added goods have much potential.

Autumn Olive (*Elaeagnus umbellata*) and Goumi (*Elaeagnus multiflora*)

These Eleagnus species produce edible fruit. Autumn olive ripens its tiny fruits in October and is considered highly invasive in many regions.

Goumi ripens its slightly larger and better fruits in spring. Both are nitrogen-fixing and easy to grow.

Future Market outlook: The tiny berries have little marketing potential, but Goumi could possibly be utilized as a niche berry crop.

Bush and Nanking Cherries

In 1999 the University of Saskatchewan released its first true bush cherry, called 'Carmine Jewel.'[2] This was the first of a number of similar releases dubbed the "Romance Series:" 'Cupid,' 'Romeo,' 'Juliet,' etc. These are unique, important horticultural achievements that took over 40 years of breeding and development to accomplish.

We've been trialing them on our farm for a number of years and my feeling is that they probably are best suited to colder areas north of Kentucky. However, they do have a lot of potential. 'Nanking' and other older bush cherries ('Joel', 'Jan', etc.) are too small and soft for practical marketing purposes. They are fine for backyard use and are certainly reliable. Value added goods potential.

Future Market outlook: Slim for Nanking cherry, but there is a market for the plants themselves. The newer "Romance Series" bush cherries have *great potential* for growers in northern areas and prairie areas.

Kiwiberry (*Actinidia arguta*)

Almost making it into this book, Actinia arguta, the small, fuzzless cousin of the fuzzy kiwi, is a wonderful and delicious fruit well suited to organic production, but suffers many deep challenges to becoming a viable small fruit crop anywhere outside the PNW. I attempted it myself for years and finally gave up on them. First, the vines require extensive trellising and take 5–8 years to get into substantial production, which is quite slow and would be difficult for many growers.

Second, kiwiberry vines tend to bud out *very early* in spring as soon as the weather turns warm, wherein the new shoots can get very damaged by any amount of late frosts, and not yield any crop. This is a *major disadvantage* in these days of fluctuating and unpredictable spring weather and increasing late freezes, and therefore breeders should try to select and breed for *late-blooming or delayed budding* in future cultivars. The early budding tendency makes kiwi berry *dangerously susceptible to late frosts and very risky to grow.*

Future Market outlook: The major hurdles it faces make kiwiberry production very questionable and expensive. However, some growers in the PNW and one large grower in PA (Kiwi Korners farm) are currently producing the fruit. Late spring frosts have been hampering production in PA.

Cactus Fruits

The best known and most marketable are the prickly pear cacti or *nopal.* There are both a Western and Eastern US species. The only one worth considering is the much larger and tastier Western species.

The tasty fruits (*tuna*) as well as the edible pads (*nopal*) are popular in the Latino community and could become a good niche crop in areas with a Latino population. They are easy to grow and nearly pest-free.

Future Market outlook: Decent in arid regions and in Latino and organic food markets. A small market exists for plants/pads for other growers.

Cornelian Cherry (*Cornus mas*)

This dogwood relative is popular in parts of Russia and Ukraine for processing into jams, etc. The small red or yellow oblong fruits grow on an easy to grow upright shrub and need cross pollination.

Future Market Outlook: Slim, due to small fruit size, obscurity, and tart flavor. Possible for value-added goods or in Eastern European ethnic areas.

Hardy Passionfruit (*Passiflora incarnata*)

The native North American species has much potential. Already people are familiar with the tropical passionfruit, and superior strains approach the quality of the tropical species. Quality varies a lot amongst wild strains, but better ones, including one we maintain, taste better and are the size of the popular tropical species—the size of a large chicken egg. The vine needs trellising but grows very easily and is mostly pest and disease free, and hardy to zone 5. The vine dies back to the ground in temperate regions in autumn, but comes back strong in spring and bears a heavy crop by September–October. Also has medicinal uses as a relaxant.

Future Market Outlook: With some breeding work, possibly even hybridizing with tastier tropical species, this species could easily become a new crop.

Elderberry (*Sambucus canadensis*)

Easy to grow and very productive of healthful berries. It's hard to imagine these panning out economically on most small farms. This is due to a few factors. First, the berries are tedious to harvest and process. Second, fresh fruit sales for this fruit are just about non-existent. The fruit is virtually only sold as a high-priced dried fruit for herbal medicine products, and as elderberry syrup. Perhaps it is also sold as frozen berries occasionally. Nowadays, the big growers mow down their plants every autumn and so next season they reemerge and yield berries on "primocane" shoots.

Blue Fruit Farm in Minnesota, USA is one of the few small farms profitably growing elderberries on a reasonably small scale. The berries are anti-viral and have proven to reduce the recovery time for colds and flus.[3]

Future Market outlook: There is a strong market for the dried or processed fruit, which is being produced in wholesale quantities in a few areas, including parts of Missouri. Small scale value-added goods are possible and certainly in demand.

Feijoa (*Acca sellowiana*)

Now here is a real winner. Native to sub-tropical Central American highlands, Feijoa (pronounced Feh-Hoe-Uh) is related to tropical guava, but is cold hardy to zone 8–9. Some claim zone 7 but that is too risky.

This wonderful fruit is sold and in demand in areas where it grows, such as California, the PNW, and other mild winter climates. It suffers few to zero insect and disease problems. Seedlings with small fruit are sold for landscaping purposes, named, large fruited cultivars exist for fruit production.

Future Market outlook: Strong in mild winter areas with urban markets that would buy the fresh fruit. Should be considered a good small fruit crop where it can be grown.

Goji (*Lycium barbarum*)

Here is another plant that just won't likely pan out for any market grower. A better option would be to propagate and sell the plants via root cuttings. Selling the berries dried would be tedious and not very profitable on a small scale.

Future Market outlook: Slim, due to cheap Chinese imports. There is a market for the plants.

Hybrid Cane Berries

Tayberry, loganberry, marionberry, boysenberry; these are some of the most popular of the raspberry x blackberry hybrid crosses. Sometimes other similar berry species, such as dewberry (a creeping blackberry species) and others are mixed in. The fruit quality on most of them is outstanding. However, they are best grown in the mild West Coast areas. *Don't plan on growing any of these unless they are proven to perform in your area!* I have trialed many of them and they just don't perform well in much of this region or other hot, humid regions.

Future Market outlook: Very strong in areas where these berries thrive, such as in the Pacific Northwest. These fail in many other colder, wetter, or more extreme weather areas.

Rosehips

Typically, these are sold as dried fruit for herbal teas. It's hard for me to imagine this crop being profitable for the market grower. Perhaps if you were dedicated to growing the productive *Rosa canina* species specifically for the medicinals market, and you did not mind the colossal harvesting and processing effort it would take, it could pan out. It is easy to grow organically. Potentially invasive.

Future Market outlook: Slim due to cheap imports. There is a market for the plants.

Seaberry/sea buckthorn (*Hippophae rhamnoides*)

This nitrogen-fixing small shrub matures to about 6–10' tall and wide. The fruit is very nutritious and attractive, being bright orange when ripe, with a flavor similar to orange juice. Male and female cultivars are needed for cross pollination. Grown in Russia and Germany. The German cultivars might be more compatible with USA climate zones. Grows well in mild summer areas only.

In the recent past I saw seaberry juice selling at a local health food store for $30/liter. So, it's working for someone out there.

Future Market outlook: Slim due to cheap imports. Co-ops could form to grow the berries for mass juice production. There is a market for the plants.

Munson Grapes

Thomas Munson (1843–1913) was an avid viticulturist who became one of the leading experts in native American grapes, and his studies were instrumental in saving the European grape and wine industry from disaster during the late nineteenth century, due to a number of serious pests. He is also credited with creating over 300 new cultivars, making him a *plant hero*! A number of these are still available, particularly through Grayson College in Texas. You can obtain free Munson grape cuttings from Grayson College.[4] Many growers are finding these grapes easier to grow organically and adapted to many different areas, including alkaline soil and the heat of Texas where most grapes fail. These are worth looking into as demand for organic grapes continues to rise. Many regions will have success with these where most other common types fail to thrive.

Future Market outlook: Excellent if you can dedicate to it and can convince people to buy seeded grapes. Wineries could provide a market.

Che (*Maclura tricuspidata*)

In my experience of growing che, it is a neat looking, low-maintenance fruit but the flavor is really mild and bland. They're a hot item in the nursery trade because people repeat claims that they're great, but most of those people have never eaten or grown the fruit. If plant breeders developed the fruit to be sweeter and more flavorful, and perhaps to ripen a little earlier than late October, it could become a very good crop. As it is now, it's very mediocre eating and not marketable.

Future Market outlook: Slim due to currently low fruit quality. There is a market for the plants.

Honeyberry (*Lonicera caerulea*)

This one almost made it into the main section. This curious cousin to Japanese honeysuckle has potential for people in cold winter areas with mild summers, such as the Midwest, Canada, and New England areas. Essentially, they are a species of upright growing non-invasive honeysuckle bush that produces tasty and nutritious edible blue berries. They originate in very cold northern areas like Russia and Siberia. The fruits are small, deep blue berries about the size of blueberries. Overall, the flavor is fruity, tangy/sweet, and good ones are delicious. Some types in some areas taste bitter. They are easier to grow than blueberries and are not as finicky about soil type and pH. They grow rapidly and need little care beyond annual pruning, fertilizing, and weed control. You need cross pollination by planting at least 2 different cultivars. They cannot be successfully grown in intensely hot and humid summer regions like much of the south. On our farm they get about ½ day full sun. Note: in warmer climates they can defoliate in the hot summer weather but are undamaged (apparently).

The bushes are very cold hardy (zones 3–4), need minimal pruning, and have very few pests. The harvest is complete well before SWD appear. You need to have at least two compatible varieties for proper pollination and good yields. I'm a fan.

Future Market outlook: These have potential in northern and prairie regions but lots of customer education is needed to sell the fruit. There is a market for the plants.

PART 3

Harvesting, Marketing, and the Future

22

Harvesting and Post-Harvesting

When handling food for the public, hygiene and food safety are a must. Many laws and regulations now dictate some of these basic rules, such as GAP (Good Agricultural Practices), etc. If you plan on selling, you must familiarize yourself with the current rules and regulations and may need to submit applications or be inspected, so research and abide by both local and national rules.

Good hygiene means showering before you go to harvest produce, and washing your hands with disinfectant soap and water thoroughly before picking. Hand sanitizer's not a bad idea too. Early morning, before the sun is hot, is the very best time to harvest any fruit. In the morning the fruit is somewhat chilled, slightly moist, highest in *prana*,[1] and low in *field heat*. Field heat is a strange phenomenon, wherein during storage the sugars and carbs begin to break down and produce heat that makes the product rapidly begin to spoil. Rapid chilling (through refrigeration) of the produce reduces field heat.

Insect and bird damage also peaks by mid-day. If you harvest blackberries in mid-day you'll usually be shaking off beetles and shooing away birds. If you harvest the ripest fruit every morning, the patch will (1) be less attractive to pests that are primarily attracted to dripping ripe fruit and (2) less damaged by said pests because you beat many of them to it. I've noticed a big difference with this simple strategy.

Harvest when fruit is ripe, but be aware that insects and animals can damage very ripe fruit, and it may be too soft to make it to market (as with figs) so learn what stage of ripeness is appropriate. It should be sweetened up but not so soft ripe that it turns to mush. Harvesting blackberries, raspberries, tomatoes, and figs just shy of peak ripeness is best.

Harvest efficiency

Get fast at harvesting! It's always most efficient to harvest with *both hands.* This necessitates that your harvest containers are strapped on at your belly and the collapsible crates lined with quart containers (mentioned earlier) are on the ground or cart. You absolutely should not be holding a container in one hand and harvesting with the other. This slows you down 50%! Both hands on deck.

Credit: Eli Stoltman

Both hands are efficiently used when harvesting fruit. A Johnny's harvest "bucket" strapped to the back allows hand-free usage. Three quarts fit inside a tub, so you can harvest directly into the quarts. Then the quarts can go into a collapsible harvest crate.

If you harvest directly into quarts or pint containers, then you don't risk crushing any berries or soft fruits in deep containers, and also you *bypass* the added labor of taking back harvest containers, picking through them and individually packing them into marketable containers. You've saved a whole laborious step. Just place the pre-packed quarts or pints into the collapsible harvest crates. The crates can then be easily stacked and hauled away to cold storage or a market truck. *Always aim to identify how you can reduce and eliminate in-between steps, from planting to the final selling.* Make sure your packing containers don't damage the fruit, and don't pack product too tight and cause it crush and leak juice.

It's usually best not to "stash" multiple berries in your hands while picking but to hold no more than one or two before getting them in the container. They can easily get crushed or dropped. Make sure you are lifting branches, leaves, and stems to locate all ripe berries, and check both sides or you'll miss a lot of fruit. Keep your fingernails trimmed short so as not to cut the fruit. Soft gloves might be useful.

Harvesting may require bending and squatting. If that is difficult for you, dedicate to doing very regular yoga asanas for the back and legs, stretching, and perhaps looking into wheeled garden chairs, or hiring help, or getting the whole family or friends involved. If you run a small farm could trained WWOOF volunteers or employees do the harvesting? Pruning for shorter (or taller) plants may help also.

Protecting, storing, and chilling

Make sure harvested fruit stays out of the sun! It's best to harvest predawn or right around sunrise, to minimize field heat and sun exposure, and your own stress. Keep it covered, or in the shade. As soon as possible after harvest, either get it directly to market within a few hours or chill it, or both. You will eventually need a walk-in cooler installed if you're doing this on any scale—see Chapter 8: Tools of the Trade. Keep the fruit chilled at 34–39°F (1–4°C). Make sure it does not freeze!

Chilling the fruit also eliminates most issues with *Spotted Wing Drosophila* (SWD). Quick chilling eliminates *field heat* and makes the fruit keep longer. Ideally, very perishable fruits like these get picked no more than one or two days before market, are chilled, and handled very carefully. This will require multiple market days or multiple marketing avenues (restaurants, health food stores, etc.) Schedule those deliveries around harvest times and market day. Unless you're selling the fruit within a few hours, it has to be chilled. However, avoid chilling tomatoes if possible, especially for farmer's market sales, as it ruins their flavor. Keep them in a cool, indoor location.

If berries have white drupelets from stinkbug damage, maybe a few here or there are passable, but it should not be obvious. The berries cannot be torn open or damaged, bird pecked, or leaky. Anything like this turns into a mold factory in the cooler. Moldy berries will upset

Stinkbug feeding damage can cause unsightly white drupelets like these.

customers. If you damage or break a berry, or if a berry is deformed, half-ripe, or unsightly, put it into a value-added goods harvest bucket to be made into jams, etc.

If you are unable to pack directly into clamshells or quarts for some reason, or the crop necessitates cleaning, transport the crates to an indoor, hygienic packing room, quickly sort through the berries (if you upscale, *berry cleaning machines*[2] will greatly speed up the work) and pack them into ½ pints, pints, quarts, etc., then pack these into collapsible harvest crates, and immediately chill to 34–39°F (1–4°C) or the specific temperature for your harvest.

Benefits of pre-packing and chilling

1. The work is completed, the product ready to go, and come market day, you won't be spending hours sorting and packing fruit.
2. Any berries that mold will ruin *only* that single container they are packed in and not a larger container.
3. The berries won't sit too deep in harvest containers and crush the bottom layer.

Finally, you can easily move the stacks of collapsible crates full of pints and quarts of fruit and pack them into your market vehicle.

23

Market Planning and Strategies

If one thing will define the successful small grower of the 21st Century it will be their ability to strategically market their product to achieve maximum net profit. Marketing is mandatory for the resilient 21st Century small grower. *Usually this will mean direct marketing.* For those of you with no marketing savvy or marketing knowledge, *it's time to get with it*. The *route* by which you market your goods is the first and most vital decision. Next is the *method*. Third are the finer details we will discuss in this chapter.

You have *two main routes: wholesale and direct marketing.* Wholesale means just that: you sell the whole lot—or a lot at once, to one buyer. For instance, you harvest 1000 pounds of muscadine grapes and choose to sell them all at one time to one buyer (such as an organic food co-op). It's risky, challenging to arrange properly, and you make drastically less profit. Only go this route if you have to. However, it is easy, very fast, and convenient for the grower, as someone else is responsible for the retail marketing side.

Direct marketing is selling direct to the customer themselves. You get a higher price for your products, and also accomplish a number of crucial goals all at once: exposure of your farm business, creating a customer base, generating customer loyalty, and getting the *maximum price for your products.* For instance, in 2021, *wholesale*, conventional muscadine grapes sell for around $2–4 per pound, while they *retail* online for $10–12 per pound. That's potentially 250–500% more profit per pound!

Direct marketing is highly advantageous, because instead of relying fully on one or two wholesale buyers who may dump you, or force you to lower your price at any moment, you instead have dozens or hundreds or thousands of loyal retail customers, and *you* choose the price! That's a real safety net for your farm.

Farms operating by this direct-marketing system rarely crash, whereas conventional farms relying on wholesale markets and distributors crash all the time. Direct marketing takes more work and effort, but if you got into small farming it's likely that you were not looking for something easy, but a meaningful, regenerative, and adventurous career or side gig. Let's go into the methods of retailing your products.

Fruit is hot but be sensitive to region

If you are currently a successful market farmer, your farmers market customers will want to buy your organic local fruit, no question. They will enthusiastically request it! You won't be able to grow enough to meet the demand once people catch on. As long as it's high quality and fresh, it will fly. Consistency and quality will be what matters, and a reasonable but confident price tag.

Once you know a product has local demand, you need to research *pricing* for the product. What to charge for a pound, a pint, a ½ pint, a quart, a bushel? Check local grocery stores and markets. Look online. When selling my produce, I make an effort to *not* undercut the grocery store. If our product is fresher and higher quality than what the grocery stores sell, why would I sell it for cheaper? At least the same price is usually fair enough, sometimes not. You will have to educate and convince your customers of your superior quality. This can be done through intelligent and convincing marketing, brochures, websites, infographics, interviews, etc. Your product is better, and here's *why*. Now your higher price is not only justifiable, it's practical and necessary. In fact, it's so good, it should be *even* higher!

Important to mention, regardless of where you sell, you should *seek opportunities to advertise* and create interest in your farm via online marketplaces: Craigslist, Facebook, Instagram, etc. If applicable, get your operation listed in official state agricultural directories, websites devoted to local agriculture, online newsletters, local newspapers, fruit growing forums, local clubs and groups, and through social media outlets. If it's a backyard gig, this is not recommended, as you could get hassled for operating an "agricultural operation in a non-ag zoned area," etc.

Containers

You'll need to acquire proper containers for marketing your products.[1] You can't just hand customers bags of berries. I once saw a newbie at a farmers market doing that, and the berries were a berry smoothie in a bag before the customers left the market. Not acceptable. Also, some containers such as wooden quarts are lovely but small berries like raspberries will slip right out of the open corners. The container has to be sized for what you're packing. Your customers don't want to pay "top dollar" for fruit and have it spilling out or turning to mush by the time they get home. They won't come back, so be professional.

Customers love options so they can choose how much is right for them, so provide a variety of size options: pints, ½ pints, quarts. Mostly you'll need clear plastic "clamshells" like you see at the grocery store. Plastic is of course, not ideal, but it is virtually the only current food-safe option, unless you go for little paper pulp containers or wooden ones (these might work for you). Encourage your customers to reuse and recycle. That makes plastic less wasteful. Biodegradable bio-plastic options may be available.

Direct marketing methods

Fresh fruit sales for a You-Pick farm

Small fruits can become an important crop to bring in more revenue and excitement to your existing farm. Many small fruits coincide nicely with the harvest of other popular You-Pick fruits: peaches (blueberries, late strawberries, and blackberries) and apples/pears (red raspberries, primocane black raspberries, late blackberries). In northern regions you might try honeyberries, currants, gooseberries, etc.

Small fruits could easily become another fresh fruit product available for customers *already walking onto your operation* looking to buy fruit. You should also harvest the fruit yourself and offer it packed in clamshells from your storefront.

In addition to fresh fruit sales, your You-Pick could offer value-added goods: berry muffins, jams, ice cream, syrup, berry plants, etc. A local You-Pick in KY makes ice cream and baked goods with black raspberries, blueberries, and other small fruits and offers these from their storefront, along with apple cider, pre-picked apples and pears, etc.

Another viable option that should not be overlooked is selling gallon bags of freshly frozen berries. Virtually any berry could be sold this way but blueberries, strawberries, and raspberries would be great choices. It's not unusual for vendors at farmers markets to sell frozen products, such as meat. You could bring a large cooler with reusable frozen ice packs and clean bags of frozen berries. I saw a vendor doing this at a market with gallon bags of local blueberries and they were flying off the table (I think they were asking $20 per gallon bag). You could sell them by gallon bag, quart bag, or by weight. Make sure to follow all local regulations, as this may be a bit of a "grey-zone."

Fresh fruit sales to restaurants and breweries

This is sometimes the most effective way of moving lots of fruit quickly and easily, especially highly perishable fruit. First, we'll discuss restaurants because the two avenues operate much differently.

Restaurants are sometimes difficult to work with; just ask any longtime local produce farmer. The reasons for this are that restaurants are high-stress environments and have a high turnover rate of employees, including chefs. Many chefs and restaurants are reliable, reasonable, and easy to work with. However, be very careful about getting your hopes high when a chef acts very excited about your products, or makes promises, and plan wisely. If you're working with a chef, check in several times over the season to stay in contact and make sure you're both on the same page (and the same chef is still there!).

Breweries are a bit different and might prove easier to work with. Local fruit wines and fruit-infused beer are becoming something of a hip gourmet item in some areas currently. Contact local breweries and also wineries and talk to the owners. Try to set something up, on the premise you're a local farmer and can provide interesting products for them. Be very communicative. Tell them exactly when you'll have fruit later in the season (go at least 2–4 months before fruit is expected). If they act favorable, try to set something up. Try to get them to agree to a certain amount you feel 100% confident you can deliver. Deliver on time.

Value-added goods

You can also process your fruit harvest into jams, wines, beers, breads, muffins, pies, etc. Fresh local jelly is surprisingly popular at farmers markets. Your A-grade premium fruits can be sold retail/wholesale while your B-grade fruits can be processed into value-added goods. It's a great use for good quality yet small or un-marketable fruit. It can make the operation more profitable. Frozen fruit is also a possibility. Don't underestimate the potential of Internet sales and marketing with value-added goods. Value added goods are generally small, shelf-stable, and easy to ship. Make sure all food safety and local laws and regulations are being followed. You may need commercial kitchen access/certification.

Farmers markets

If you *really* want to make decent money growing fruit, direct marketing at a *busy* farmers market is going to be by far the best route, as long as you have access to a reasonably attended market with customers who will pay good prices for quality products. This may necessitate traveling to nearby cities to market your product. *Ask local market managers what their average shopper counts are and go for the most populated markets you're willing to travel to.*

In most farmers markets around the USA, fresh local fruit is in very short supply. If you bring really nice fruit to the market, it is almost surely going to be an instant hit. But there are some ways to increase your likelihood of success.

First, presentation is everything. You must have a nice setup. Multiple (2–4) large, sturdy folding tables work well. Creating a U shape with the tables creates the sense of space and brings people into your area, and then helps contain (corral) them. You need a collapsible tent or canopy overhead. Don't waste your money on cheap tents unless necessary. Make sure the tents are secured to the ground! Tables need a nice table cloth (preferably stain resistant). You need a professionally printed bright vinyl sign with your farm name and a few examples of your top products written very large (ORGANIC BERRIES, ORGANIC RASPBERRIES, etc.) These can be designed and purchased online, or find a local printer. If you are Certified Organic, get the green and white USDA seal printed on the sign. Hang the sign in a visually prominent place.

Your products need nice bins to sit in. Bins make the product look orderly, abundant, and clean. Make the bins look *very* stacked and always abundant. If product is getting low, remove the bins or put into a smaller bin.

Create lots of hype by informing your customer base *weeks ahead of time* that you'll be bringing your luscious fruit to the market. Send out marketing emails, social media posts, etc., weeks before ripening and again a few days before you bring the fruit to market, and the day of market. Can your market coordinators help create some hype by messaging the customer base? (*"Fruit Jam Farm" will have fresh organic blueberries next week.*) Get your farm in the local papers! You are doing this ultimately to sell the fruits, right? You must create lots of hype!

Have a clean and *state-certified* scale handy and lots of new bags, if needed. Basket-type scales and bags are good for durable fruits like grapes, tomatoes, etc. Having some empty cardboard boxes handy is a good idea in case customers want a lot of product.

Behavior. This is a crucial art to master at the market. You want to exude positivity, alertness, and care. If you are sitting down playing on your phone, you exude a *careless and lazy vibe* and you will lose *lots* of potential sales and interest as people breeze right past your downturned gaze. *Never sit and never, ever play on your phone.* People that are going to buy, *purchase*, and don't usually have to ask lots of questions or text you five times.

Always be standing, friendly and alert. Greet everyone with a simple "Hi, how are you doing today?" or something similar. You can then make a simple sales pitch, "Our tomatoes are at their peak, and going fast!" "Muscadine grapes are just now in season, and we have ½ pound and pound bags." "This is the very last week for blueberries, can I interest you in a pint before they're gone?" You will get used to doing this as you do it regularly. It will drive sales *way* up.

When a customer wants something, after handing it to them, try to make it lead to something else by confidently using this line: "*What else for you today*?" not "Do you need anything else?" The reason being, the first question leads to a mental exploration of your products and is not answered with a quick yes or no. You could suggest something like "We also have XYZ available on sale." This will up your sales also, a lot.

Be friendly and upbeat at the market. You can be as personal or generic as you like. However, it's proven that calling your customers by name creates loyalty to your farm and ups your sales. Try to memorize and use customer's names. Greet everyone who shows up at your table. But be careful about talking too much, getting too overly charismatic or making jokes, as people can easily get offended or put off and never come back. Too much talking can also distract you from oncoming customers and make you lose lots of sales. You don't want new customers arriving to your table feeling like they are intruding on a personal conversation you're sharing with someone else. You must be present for everyone. Sometimes you might need to rapidly end a conversation with an energetic, "Hey, it's been great seeing you!" and then a quick turn to the new customer with a cheerful "Hi, how are we doing?" Remember, social interactions are fun and meaningful, but your market farm will go under if you don't rack up the numbers by interacting with customers and upping your sales. It's the reason you are there as a marketer and not a customer. Farmers markets are the time to make sales and finance your life, not primarily to hang out and socialize. Do that afterwards.

If a customer is ever displeased with your product ("My figs were moldy"), replace it for free, no questions asked. Studies have shown that this builds very strong customer loyalty, when situations arise and are dealt with nicely in their favor. You can either "give up" a $5 clamshell of berries or a repeat customer; you choose.

An important rule of thumb in your business is: be nice, courteous, friendly, and always polite. Even if you are struggling and having a hard day, act as positive as possible and *never, ever* bring your personal drama to your table. Your business should be a "*drama-free zone.*" A positive attitude will attract positive outcomes and scenarios. People like to buy products from smiling, optimistic, friendly people (and bypass the ones frowning or playing on their phones).

Online marketing

Finally, don't underestimate the potential of marketing less perishable fruits such as muscadines, as well as seeds/cuttings, etc., via online marketplaces like Etsy, Facebook Marketplace, Ebay, Amazon, Craigslist, etc. Once you get the hang of these platforms, as well as how to package your product for shipment, it can really work out well. USPS "flat rate medium boxes" are often ideal. Focus on perfect quality, good communication and descriptions, proper packaging of the product, and getting plenty of 5-star reviews through the same. Make sure ads and listings get removed or deactivated when the product runs out!

24

The Future of Small Fruit Growing

Climate change and fruit growing

Undoubtedly, climate change and extreme weather events will continue to impact the global cultivation of fruit, far more than the production of vegetables. The reason being, most vegetables are simply the fast-growing vegetative parts of annual plants (leaves, stalks, roots, etc.) Commercial fruit requires flowers born on (mostly) perennial plants. Flowers are fragile and sensitive to harsh weather, such as extreme or prolonged rains and winds that disrupt pollination, and late spring frosts that can destroy dormant flower buds, etc. Vegetable planting can be strategically delayed based on local weather. With fruit growing, flowers bloom when the plant's genetic timer tells the plant it's time to open them. This works great when conditions are stable or based on the plant's genetic history (which its internal genetic "timer" was based on) but fails when a curveball season throws a late freeze event during full bloom. Those curveballs seem to be increasing in regularity.

Flowers are genetically programmed to bloom once certain criteria are met, such as "chill hours"—when a sufficient amount of time has passed below 45°F (7°C). It's actually more complicated than that[1] but this is a good basic explanation. Native plants are genetically "programmed" to bloom at the appropriate regional time that spring arrives. But now many plants are experiencing climate change (or have been relocated around the globe) and that poses a problem that often cannot be overcome. Green and leafy, yet fruitless trees and bushes are the result. Dead trees and bushes can be the result of severe late freezes after plants have started to grow.

Climate change will be impacting many fruit growing regions, and may require the changeover of cultivars (or even whole species) and the phasing out of many common ones to newer, better adapted cultivars, or, the *revival* of hardier, more resilient older ones as well. Georgia peach growers are having to transition entire fields of peaches to low-chill cultivars as winters warm up.[2]

For another example, a few years ago I read an article about a family farm that grew apples in a hot, dry region (I believe it was Southern California). They found that as it was getting hotter in the region, apples were no longer producing fruit like they used to and were no longer commercially viable. They switched over to jujubes (*Zizyphus jujuba*) and the more resilient jujube trees flourished in the dry, intense heat and produced copious amounts of valuable fruit, saving their farm. We'll see more shifts like this, with "new" fruit being introduced and grown in new areas and fruit species "migrating" both north and south, east and west.

My proposal to all fruit breeders and fruit explorers of any level is to step up our breeding and selection to focus on the following crucial goals, designed to help us achieve successful and resilient fruit growing:

1. *Late Blooming/high chill hours.* Our fruiting plants need to bloom later in the season. Staggered bloom times will also be beneficial. 'Ponca' blackberry, for example, blooms over the course of several weeks instead of several *days.* This improvement will prove critical in protecting the crop from late freezes. On that note, in general it is best to avoid planting fruit plants that are described as "early blooming," unless you live in California or the Pacific Northwest, where late freezes and frosts are currently less of a serious problem (but that may be changing also in areas of the PNW). Also, areas of the South may need specialized low-chill cultivars as winters shorten and get warmer.

2. *Hardier blooms and buds.* Select for plants that have buds that can withstand colder than normal temperatures when dormant, without getting frozen off. They must be able to withstand extreme wind chill. We need plants that can bloom and withstand some light frost and still have viable ovaries that finish their reproductive cycle to fruition. Example: 'Contender' peach can be in full bloom, experience a frost, and still set a large crop of quality peaches. Most would fail under the same conditions.
3. *Heat and drought resistance.* Some regions are getting slightly hotter and drier, with increased UV levels, and this is likely to get more severe in time. We'll need cultivars and rootstocks that don't wither up or drop their fruit in the heat. More resilient species, such as jujube will need to be emphasized and trialed. Traditional and heirloom cultivars from arid Middle Eastern areas, Mexico, India, the Southwest USA, and Australia might prove very useful in this respect, as well as for breeding purposes.
4. *Moisture, disease, and insect resistance.* With more global growers rapidly going organic, we need cultivars that can handle disease and insect pressure without resorting to toxic synthetic chemicals. And, organic or not, pest pressure is increasing for everyone due to the introduction of new invasive insects through global trade. Climate change also puts stress on plants, leading to increased susceptibility to insects, as described earlier with Asian ambrosia beetles.

   Part of this strategy may include developing cultivars with *reduced days to maturity and also earlier ripening times* to have the crop ready earlier in the season before seasonal insect populations peak. Considered advantageous for avoiding SWD, this trait is being implemented in modern bramble berry breeding. Conversely, fruits that ripen in autumn may experience reduced pests as well, with the exception of SWD.

   Moisture resistance means resistance to excessive periods of rain and humidity without it ruining the crop. For instance, ripening figs are usually ruined by rains. However, the late Dr. O'Rourke, of Louisiana State University, developed in the mid–late 1900s (often using the very resilient heirloom fig 'Celeste' as the mother tree) hundreds of hybrid fig cultivars, many of which produce useable and marketable fruit even in the extreme humidity and moisture of the Gulf Coast, USA. It can be done. As rains become more unpredictable, regions prone to heavy summer rains will need built-in moisture resistance in case summer storms turn heavy and prolonged. This is a challenging one.

5. *Increased nutritional value.* We need our produce to be as full as possible of nutrition. Plants with high levels of nutrition and *brix* often exhibit lower rates of disease and insects. By selecting for fruits that contain unusually high levels of antioxidants, minerals, calcium, phytonutrients, anthocyanins, etc., as well as improving soil quality through organic cultivation, we can help ensure a sustainable and substantial improvement in the health of people and the planet. *We have to.*

Rescuing the genetic pool by selecting your own local cultivars

This is for those interested in experimentation and trial, and also those wanting to help humanity by introducing good plant genetics to society at large. *Market growers should almost always stick with tried and true cultivars—including local cultivars that are outstanding and/or popular.*

However, experimentation on a small scale can have promising results. Just be aware that any cuttings and plants you obtain can harbor viruses, diseases, and insects that can spread to your plantings, so be very careful. That said, if you are interested in trialing and exploring promising cultivars of plants, and helping our genetic crisis, this is what I suggest:

1. *Start with what you can find locally.* If you find something promising, take and propagate some cuttings or suckers. Always collect material legally and with permission. Check with owners of local farms, managers of botanical gardens, arboretums, historic sites, historic gardens, and plant collections. I once found an incredible grove of very productive and very delicious juneberries growing in the middle of a

shopping center parking lot, and propagated them. I've made many such valuable finds. For that reason, I always have snips and bags in my car.

2. *If very little is available or found locally, then reach out onto online forums.* Some suggestions are in the Resources section. Be very clear about where you are and what you're searching for. Try to source plant materials that are likely to be adapted to your specific conditions.
3. Search through germplasm repositories, edible plant collector groups, and seed/plant exchange groups. Go through the USDA germplasm collections and make a request.[3] Check the *Resources* section at the end of this book for more links.
4. *If interested in trialing cultivars from other countries, either find them online through forums or nurseries specializing in exotic cultivars.* Beware of buying materials like cuttings and seeds randomly online or on eBay. Very often the material will be fake, mislabeled, damaged en route, dried up, or could harbor pathogens. It can also be illegal and the USDA might come knocking on your door, flashing their badges, and with boxes ready to fill with your plants. (I know someone this happened to.)

That being said, if you are looking to explore international cultivars, look to regions that are at similar latitudes and/or climate zones. For instance, KY has a similar climate to continental Europe and much of China. The Gulf Coast of the USA has a sufficiently similar summer climate, and long enough growing season, to make certain annual crops from India, Thailand, the Caribbean, and tropical Mexico possible. Canada has climate regions similar to Northern Europe, Russia, and Siberia.

Sourcing plants and cultivars from similar climatic regions might lead to breakthroughs. This is currently being done successfully in the backyard vegetable seed trade as well as the fruit nursery trade. Many excellent and superior fruit and vegetable cultivars are currently becoming available in the USA, that originated in Ukraine, Russia, Thailand, India, China, and Western Europe.

5. *Start seedlings of resilient cultivars and specimens*. That way, newly emerging hardy genetics can be identified that have adaptions to the local conditions. Remember, some plants (such as apples) require hundreds or thousands of seedlings in order to identify superior marketable stock, and some species produce male (non-fruiting) and also female (fruiting) offspring (including mulberries). However, it's worth trialing and seeing what comes of it. Keep careful notes.
6. *Trial, test, record, and openly share findings*. Just because something does "awesome" it's first or second year does not mean it is truly locally resilient or adapted. The true test is long term, and a glimpse of general adaption can only be ascertained by observation over a minimum of 4–5 years. If a season throws extremes in there, even better for testing resiliency. Keep notes and careful track of cultivars. Don't be overzealous to the point of getting cultivars lost, confused, or mislabeled. Be dutiful and patient; label and map carefully. After testing, make sure to share your findings online and appropriately share plant materials with other communities and growers.
7. *Introduce any new superior findings into the nursery trade*. This is possibly the best way to help society at large and our plant genetic crisis. Contact nurseries dealing in unusual or rare plants and let them know what you have. Provide detailed descriptions and photos. They may be interested in buying cuttings, and introducing it. This way it can get rapidly propagated and distributed to thousands of growers. This has recently happened with the development of 'Niwot' raspberry and the discovery of 'Olympian' fig, both by non-nurserymen, but introduced into the trade and now grown by thousands of people. Also, you might contact universities doing research on the species, and also donate plant material to the USDA Germplasm repository. You can contact the author about new superior findings at: blake@peacefulheritage.com

Through this 7-step process we can personally and rapidly reclaim much of our lost genetic diversity and bring resiliency back to the landscape of our communities and farms and into the hands of our children and grandchildren.

25

Interviews with Successful Small Fruit Market Growers

The following interviews are with people out there doing small farming with small fruits and making it work. As I was reading these interviews, I noted that a number of major points and strategies I suggested throughout this book are echoed and affirmed loud and clear by these successful, unconventional growers.

In order to grow and profitably market their unusual crops, they focus on these things:

1. *Education and outreach.* They reach out, intrigue, and educate the public about why their product is desirable and worth the money.
2. They usually focus on, (or at least implement some form of) *direct marketing the product and retailing it to the end user.*
3. They all focus on *small acreage* and on a multitude of diverse crops.
4. *They are creative, ambitious, patient and work very smart and strategically.* These strategies helped them all to manage their operations successfully.

Let's get into it.

The first interview is with Jim Riddle, who currently co-owns and operates Blue Fruit Farm in southeastern Minnesota with his wife Joyce. Blue Fruit Farm is a successful certified organic specialty small fruit farm in the challenging fruit growing climate of Minnesota. Here are the questions I posed to Jim and his insightful answers:

Q: What inspired you to undertake growing unusual and questionably profitable fruits such as aronia berry, honeyberry, etc.?
A: When we started Blue Fruit Farm, my wife wanted us to grow blueberries, which need a soil pH of 5–5.5. Our farm is in the Driftless Bioregion in Southeast Minnesota, and our native pH is around 6.8–7.2, so we amended the soil with high quality compost, peat moss, and elemental sulfur for the blueberries to lower the pH. I wanted to grow fruits that would thrive without changing the soil pH, so we chose black currants, elderberries, honeyberries, aronia, juneberries, and plums. We also wanted to grow fruits with high antioxidants and high levels of vitamins and bioflavonoids, and the unusual fruits we grow are much higher in these nutrients than blueberries.

Q: When reading profit forecasts for such fruits, a lot of times people will say the only way to make these profitable is by planting large acreages. For instance, I just read online that an expert said juneberries are only profitable if at least 40 acres are grown and harvested. Has this been your experience? If not, why?
A: It is really important to size your plantings to fit your potential markets. Instead of focusing on wholesale markets, we practice *relationship marketing*, where we get to know our buyers and they know that we produce high quality, organic fruit. All of our fruits are hand-harvested at peak ripeness, which means peak flavor. By providing exceptional quality, we have developed a dedicated customer base. We have found that our direct sales system is more profitable than the fruits we sell at wholesale prices in larger volumes, but it is good to have a mix of different markets, because we would not want to rely on just one form of marketing.

Q: What do you think are the major hurdles to someone creating a profitable niche with unusual fruits and berries?
A: Everyone loves blueberries! I like to say that blueberries are hard to grow, but they sell themselves. The other fruits we grow are easier to raise, but we have to sell them! We have done numerous presentations in the area, always offering taste tests of the jams and juices we make,

to introduce potential customers to these new fruits, which are packed with flavor.

One major hurdle is the fact that black currants were not allowed to be grown in the United States until recently, so American customers are not familiar with their bold, tart, musky, and smoky taste. In addition, small dried Corinth grapes are allowed to be sold as "dried currants," when they are really just small raisins. People think they have eaten currants, when they have really eaten tiny raisins!

To be profitable with unusual small fruits, plan on spending quite a bit of time *educating your buyers* about the fruits. Offer taste tests and recipes. Make it easy for people to buy and use them. We have put a lot of effort into educating customers about black currants, aronia berries, honeyberries, and the other unusual fruits we grow, and people have really responded. It is gratifying to watch as people taste new fruits, expand their dietary horizons, become regular customers, and help spread the word to their friends and families.

Q: What would you estimate are your sales percentages for fresh fruit versus value added goods versus frozen fruit? (ex: 25% fresh, 25% value added, 50% frozen).
A: We sell approximately 40% of our fruit fresh; 50% frozen; and 10% as jams and juices. At the same time, about ⅔ of our sales are wholesale and ⅓ retail, but retail sales have increased with the addition of You-Pick days at our farm.

Q: What has surprised you along the line with this farming endeavor?
A: We did not anticipate the *bird pressure* and the large amount of fruit that we would lose to the birds, early on. We already had an 8-foot-high fence around the 5-acre field to protect the fruit bushes and trees from deer, and we knew we would have to protect against raccoons, which we do using an electric fence and live traps.

Q: Can you tell us in detail about how you protect your crops from birds.
A: We employ several strategies to protect our fruit crops from birds.

The most effective is our Smart Net system. We installed untreated wood posts and wires to suspend bird netting nine feet above the ground. There are side curtains around the field. The suspended netting allows us to operate equipment and harvest fruit. It also allows us to let the fruit reach peak maturity and flavor before we harvest. In addition to the bird netting, we operate a bird guard sound device, which emits cries of distressed birds and bird-eating raptors during daylight hours. In 2021, we installed a laser bird deterrence system to scare birds from feeding on our fruit. We currently use the laser system in combination with the overhead netting.

Q: What have been the most successful crops and what have been the most challenging or disappointing?
A: Blueberries are quite popular, but it took about 6 years before the bushes started to produce marketable yields. Fortunately, black currants, elderberries, aronia and honeyberries produce fruit about 3 years after planting. Elderberries have a strong market demand, but there are several production challenges, and the fruit must be de-stemmed before being bagged and frozen for sale. We have been disappointed with 'Night Mist' and 'Midnight Blue' honeyberries, which were so bitter that we couldn't even make them into good jam, so we removed those two varieties.

Q: How did/do you get customers interested in these unusual crops?
A: We have hosted a number of field days and farm tours, including our annual Blue Fruit Fest. We have given taste samples at our local food co-op and farmers market and other local events. We have worked with chefs to introduce these unusual fruits into restaurant menus. We also worked with the Minnesota Farmers Union, who had aronia/blueberry ice cream bars made with our fruits, with the ice cream bars being sold at the Minnesota State Fair. We've been interviewed by newspapers, magazines, and television stations where we described our unusual fruits and how they can be used in common foods, such as smoothies, muffins and scones. We give out recipes. Finally, we're active on social media and keep people posted on farm activities, even during the winter.

Q: Which crop is your personal favorite to work with and why?
A: My personal favorite fruit is the honeyberry. The honeyberry, or haskap, known as the "fruit of long life," has the highest levels of antioxidants of any fruit grown in North America. These plants have been grown in Siberia and Hokkaido, Japan, for thousands of years and are hardy to USDA Zone 3. They bloom in mid-April (in Minnesota) and provide early forage to native pollinators, including over-wintering queen bumblebees.

They are packed with flavor and ripen for us around June 1st, about 2 weeks earlier than strawberries. They make incredible jam, and mix well with other fruits.

Q: Have there been any crops you have explored, that you think could also be highly successful, that you do not currently grow?
A: We looked into growing hazelnuts, but were concerned about the eight years it would take until first harvest, and rodent problems. I do think that hazelnuts have tremendous potential for some situations.

Juneberries, or saskatoons, are very sweet and have good market potential. There are some high yielding juneberry varieties, such as 'Thiessen' and 'Smoky', that are better than the 'Regent' variety we grow. If we were further south, I would look into growing pawpaws (*Asimina triloba*).

More research needs to be done to develop high-yielding, uniform-ripening varieties of elderberries that have the potential for mechanical harvest. Elderberries have natural anti-viral compounds and are proven to be effective against flu and cold viruses.

Q: Acquiring land seems to be a major hurdle for young aspiring new farmers, what with land prices being so high, etc. Any advice on this?
A: We are in the process of transitioning our operation to the next generation. It took some time, but we found a couple who are committed to organic practices and to growing healthy fruits. We plan on leasing the operation and giving the new operators all management responsibilities, while we retain ownership during the transition process. If all goes as planned, they will purchase the farm on contract. We have structured the

arrangement to maximize their chances of success, so that they are not overwhelmed by debt.

When planting fruits and other long-term perennial crops, you need to be prepared to "float" the operation until the bushes and trees produce marketable yields. You need to maintain an off-farm source of income, or else maintain annual crops or livestock to produce income, until the fruits carry their own weight. Use that time to build management skills, install a walk-in cooler and other processing equipment, and develop markets. For instance, we purchased a berry cleaner several years ago, which reduced labor time to clean debris and leaves from harvested berries prior to packing.

Q: What are your favorite cultivars of the fruits you grow?
A: The cultivars which have performed the best for us in our region are:
Blueberries: 'Patriot', 'Polaris', 'Blue Crop', 'Nelson'
Black currants: 'Titania', 'Minaj Smyriou', 'Crusader' (all powdery mildew resistant)
Elderberries: 'Bob Gordon'
Honeyberries: 'Aurora', 'Cinderella'
Plums: 'Northern Blue', 'Todd'

Q: What are your average yields on a mature honeyberry bush, and what size do they grow to, (or what size do you keep them)?
A: A mature honeyberry bush will be about six feet tall and six feet wide. We prune them to be more upright and open, allowing for sunlight penetration and good airflow. This allows the fruit to get larger, ripen fully and it makes hand harvest easier. We are getting 6–8 pounds of berries from our older bushes nearing maturity.

Q: What are the biggest disease pressures you have in your orchard, and how do you deal with them?
A: We don't really have any disease issues in the blueberries, honeyberries, aronia or juneberries. Stylet oil helps with powdery mildew.

Q: You use pyramid traps in your orchard. What is a pyramid trap?
A: Also known as Teddar traps, pyramid traps are upright black structures that attract plum curculio (PC) adults, who see them as tree trunks. The insects climb up the pyramids, and are then trapped in translucent cylinders, where they can be monitored and killed. We use the traps to monitor PC activity, and to a lesser extent, for control.

Q: Can you tell us about your rainwater catchment system and how you use it for irrigation purposes?
A: We have a 30' × 40' equipment shed in the middle of our fruit orchard. The building has gutters, which collect and feed rainwater through a mesh filter and into a 1500 gallon above-ground poly tank. From there, the water flows into an underground 4500-gallon concrete tank. We use solar power to charge deep cycle batteries, which provide power for the 12-volt irrigation pump. We have a Netafilm drip irrigation system. We can inject fertilizer through the irrigation system with a 3-gallon EZ Flow injector. When the rainwater system is dry, we can pump water into the storage tanks from our deep well.

Q: Any other final thoughts on growing unusual fruit crops or small organic farming in general?
A: When converting land to fruit production, take time to understand the fertility needs of the fruits and build soil health using cover crops, compost and micro-nutrient amendments. Get the weeds under control. Time spent building soil on the front end will pay dividends in the long run.

Take steps to provide food sources and habitat for beneficial insects, including pollinators and predators. Most of the fruits we grow rely on native pollinators, so we planted a number of native flowering plants, such as New Jersey Tea and hairy mountain mint, to provide alternative food sources. The land where Blue Fruit Farm is located has been managed organically since 1975 and has tremendous biological diversity, which really helps with pollination and pest control.

Research and experiment with new ideas for pest and disease control.

One year we had a badger set up home in our shed in the field. A friend, who was a US Fish and Wildlife Service naturalist, told us to turn on talk radio, as badgers do not like the sound of human voices. So, we did. The badgers were gone in a week, and we have not had any problems since.

Remember that organic production requires *proactive management strategies*, beginning with healthy soil. Healthy soil produces healthy crops packed with flavor, healthy animals, and a healthy planet. Customers can really tell the difference, and they will keep coming back for more!

Jim Riddle
Blue Fruit Farm, www.bluefruitfarm.com
Winona, MN

Next is an interview with Guy Ames, Horticulturist for the National Center for Appropriate Technology, (NCAT). NCAT is an organization that researches and helps implement sustainable agriculture solutions and creates many very useful publications on organic farming. Their website is https://www.ncat.org/. I highly recommend checking out what they offer.

Guy lives in the Ozark area of Northwest Arkansas and has been organically growing unusual species and hardy cultivars of fruit for over 50 years! His challenging bioregion is very hot and fairly humid, with hot summers and moderately cold winters. He is in USDA zone 7b-8. It's not an easy region to grow many fruits, but many small fruits do quite well, as we will see below.

Q: Thanks for your time today. So, Guy, how long have you been growing fruit?

A: I've been growing fruit seriously since 1980 (though started in 1972). My first planting all died in short order but I learned not to plant orchards on bottom land, even though the soil looked great. After a stint in the U.S. Navy, in 1976 I planted more fruit plants at a different site, all in accordance with various Rodale books and Mother Earth News articles. As I watched most of them succumb to various problems, I decided the books and I didn't know enough, so I enlisted as a student at the University of Arkansas, mostly to get the botanical, entomological, and phytopathological basics I didn't have. *I much prefer knowing the basics and coming up with answers for my particular problems over the prescriptions of well-intentioned farm & garden writers who are farming and gardening somewhere else.* [Emphasis added.]

Q: What small fruits and berries are you successfully growing in Northwest Arkansas?
A: Juneberries (*Amelanchier canadensis*), muscadine grapes, a few of the old Munson grape varieties, cudrang [che fruit], mulberries, blackberries, elderberries, gojis, and a few gooseberries.

Q: What are your biggest pest and disease issues?
A: *Black rot* would be (and was at one time) my biggest problem on *grapes*, but I finally found the right disease-resistant Munson varieties and muscadines to side step that one. *Japanese beetles* are still a big problem on the bunch grapes, but not the muscadines! *Stink bugs* are a huge problem on the blackberries and to a lesser degree, the elderberries. *Spotted Wing Drosophila* (SWD) is a diminishing problem on my blackberries. Birds are problematic on the mulberries, elderberries, juneberries, and bunch grapes—but not the muscadines so much!

Q: Are there any small fruits you have tried, only to give up on in your area, and why?
A: Raspberries and currants, both because of our summer heat. If I give them a lot of shade, they do better but yield less. I don't mess much with gooseberries anymore because of the difficulty of picking them and the low prices I got for the fruit. People here in the South didn't really know or care about gooseberries, in my experience, but maybe I gave up too soon.

Q: What lesser-known small fruits do you think have the most market potential?

A: First, juneberries. But make sure you have the *right species* of *Amelanchier* for your region. I messed around with *A. alnifolia* (saskatoons) too long and could never coax healthy plants and good production from them...and then they started getting cedar apple rust in long, wet springs. The species *A. canadensis* is highly resistant to rust and produces like crazy for me, so now I focus on that species.

Next, mulberries. The challenge with them, of course, is their tenderness and perishability.

Then there's muscadines! These have the most potential popularity where muscadines are historically grown and available, but here, just outside their historical range, they grow great, and are incredibly easy to sell.

Finally, elderberries have still not met, I think, their production potential. The market is growing like crazy and I don't think production is keeping up, at least in our area.

Q: What hurdles do you think exist to making these small fruits popular?

A: Education and exposure. When I bring muscadines and juneberries to the local farmers market and offer a free berry, people are almost always blown away! Mulberries are another example, but I'm not sure how to best address the handling and perishability problem.

Q: What are your favorite cultivars for these fruits?

A: *Muscadines*: 'Sugargate', 'Summit', 'Cowart', 'Nesbitt' are all excellent and do very well for me.

Mulberries: only 'Illinois Everbearing' will do, at least in my experience. I've tried 'Shangri-La' and 'Pakistan' mulberry trees in the past. And although both were rated hardy to zone 6, all trees of both varieties eventually died of cold damage.

Juneberries: get the right species for your bioregion, but, after that, seedlings might be as good as named cultivars. I have 'Prince William' (*A. canadensis*) and 'Princess Diana' (usually listed as *A. X grandiflora*) and I cannot tell any appreciable difference in height, performance, berry appearance or flavor. In fact, I can't see much difference between these two cultivars and seedlings from the fruit of these cultivars. Consequently, I no longer bother to take cuttings for clones; I just grow seedlings.

My juneberries seem to max out around 12' for me. *However*, I think I might start topping them in the future to facilitate *covering them with bird netting, as I have to do*, as the fruit ripens and starts attracting birds.

Elderberries: The varieties coming out of University of Missouri (like 'Bob Gordon' and 'Wyldewood') really do have vast superiority here over the older varieties from New York and Nova Scotia ('Nova', 'York', etc.)

Q: Do you grow figs for marketing?
A: I've given up on figs. I've grown all the usual supposedly hardy varieties, and they've all failed. I didn't try growing them in a high tunnel when I had one.

The final interview is with Duane Hebert. His small farm operation Edge of Nowhere Farm is located outside Phoenix, Arizona, which experiences extreme summer heat and dryness, and mild winters. We're focusing our questions on his cultivation of mulberry fruit, which he plans on marketing very soon.

Q: In Phoenix, Arizona, how are you growing mulberries—in terms of spacing and pruning and favorite cultivars?
A: Here in the desert of Arizona we have a near ideal climate for most, if not all, mulberry varieties. Our favorites are 'Shangri La', 'Illinois Everbearing' and 'Black Pakistan'. Of those 3, 'Black Pakistan' would be the most productive in overall yield. Spacing depends on the variety, so we range anywhere from as little as 10' on the dwarfing varieties to as much as 30' on the larger specimens. We prune all of our mulberries during the

winter dormant season and tend to be fairly aggressive with our pruning cuts. This is to ensure we control the tree size to make harvesting more manageable.

Q: Have you trialed methods of growing mulberries that you found ineffective? Any cultivars you have eliminated?
A: We have used closer spacing (as little as 6 feet), but found that the trees grow too rapidly to reasonably control the size during the growing season. There are a few cultivars that we no longer grow for production, including both 'White Persian' and 'White Pakistan' varieties. The fruit quality and taste were not as desirable as the darker fruiting varieties.

Q: Any pest issues? Do you net for birds?
A: Birds. We typically don't net for birds as our mulberries tend to produce heavily enough that we can still take a harvest even with the bird pressure. The fruit ripens very rapidly, so we also harvest both in the morning and in the evening to attempt to keep the ripe fruit from the birds. Netting can be very difficult with our two highest yielding varieties ('Shangri La' and 'Black Pakistan') as they grow very rapidly during the fruiting season. Netting would need to be adjusted every few days to compensate for this, so determining a labor cost benefit is proving difficult for us. We may revisit this once we start producing for the general public.

Q: Any major setbacks to mulberries as a crop in the Southwest?
A: The marketing aspect is challenging. Mulberries are very difficult to store when ripe and our heat (these are ripe during the hottest parts of the year) makes that even more difficult. However, they are a wonderful tree for the Desert Southwest.

26

Conclusion

It is April 2022 as I write this conclusion, and once again volatile spring weather in our region is endangering the growing of many fruit tree crops. Once again, the small fruits appear unharmed and are set to crop nicely, as they all do almost every season. Their resiliency and adaption to difficult, changing climatic conditions is well documented by now and clearly observable.

We still retain a small fragment of our fruit growing heritage through nurseries and growers that maintain older cultivars or collections. However, a large portion of the fruit cultivars we have access to now were developed through fairly recent breeding programs, developed in the age of full reliance on toxic chemicals as the solution to all pest and disease problems. This makes many of our available cultivars *very challenging if not nearly impossible to grow organically*, due to lack of basic genetic resiliency to common maladies and insects. These cultivars were never selected with that goal in mind because the liberal use of agricultural chemicals made it seem unnecessary. Resiliency was basically *bred out* in favor of highlighting traits conducive to superficial concerns of the market, such as physical resiliency to being banged around in a box on a truck, or looking shiny in a supermarket display. The majority of older cultivars we have might be heirlooms from yesteryear that originate from all over the planet, are region-specific and many are not always widely adapted to thrive just anywhere they are planted.

Optimistically, as of the early 21st century we are on the cusp of a bright new age in plant breeding and genetic diversification. Many breeders, including major universities and the USDA, have now come to realize the truth and importance of what the organic movement has been screaming for well over 50 years. We need a transition to organic agriculture, which includes reducing dependence on toxic agricultural chemical use by breeding and utilizing plants and cultivars that are hardy and naturally resilient to diseases, insects, and other stresses. Increasingly, this includes the need for cultivars that can handle extreme climatic events, such as late frosts, increased heat, increased UV intensity, shorter winter chill periods, and precipitation irregularity (excess rain or droughts).

It's becoming more common for new cultivars to be selected for better built-in resiliency, as well as superior flavor and nutrition, while producing good yields and high quality fruit. It's now becoming a win-win, instead of a short-term industrial "win" for market looks and production and a long term "lose" for humanity and the planet via proliferation of lower nutrition and weak genetics dependent on chemical inputs to yield.

There is a huge scope for a revival of resiliency and local self-sufficiency in fruit (and all food crop) production everywhere as more and more households, urban farmers and communities choose food production over lawns and parking lots. If this is you, keep it up, on any scale you can, and teach others.

One of the obstacles in this effort are corporate interests such as 'big box' stores and others selling inferior, weakened, or regionally-inappropriate fruiting plants and cultivars that perform poorly. Also hampering this effort are unscrupulous online nurseries that sell falsely advertized or inferior plants, regardless of their true pedigree, what to speak of resiliency, local adaption, health, vigor, variety, species, etc. These poor yet prevalent business practices lead to millions of dollars in utter failures and also discouragement in the cultivation of local fruit and local resiliency.

Another area for improvement is more accurate and reliable information, although that continues to improve with time and better resources such as this and other quality books, online grower forums, and the rapidly increasing number of organic-focused university research trials and subsequent scientific publications. Also needed are more high-quality nurseries and plant distribution networks.

Demand for high quality and locally produced organic food is only going to increase and become more of an urgent necessity as time moves on in the 21st century and beyond. The supply of small fruits through small, localized production represents a strategic point that enterprising and hardworking local growers can fulfill, and the niche is very wide open. By utilizing a little space, patience, resources, and hard work you can pull it off and make your land (or someone else's) and your local community more fruitful and abundant, and create loads of valuable, high demand products. At the same time we can be safeguarding, sharing, and propagating the invaluable and abundant, yet greatly diminished genetic resources we have graciously inherited from our ancestors, for many generations to come; resources that will be crucial to the survival and thriving of the human race into the distant future.

Through these means, people and communities will gradually reclaim the skills and resilient genetics needed to grow locally adapted species and cultivars of organic fruit that will give the best success in the long run. Perhaps once again fruit will burst with freshness and delicious flavors, once again backyards and alleys will abound with fruiting trees and bushes bowing low with their copious bounty, and once more children's hands will be stained with berry juice on a hot summer day.

APPENDIX
Jivamritam

To make *Jivamritam*, this is what you need:

- A clean, food grade 55 US gallon (208 liter) drum or equivalent (food grade plastic or steel drums not previously used for toxic substances work fine).
- Rainwater (preferable) or municipal water allowed to sit a day or two prior to brewing to evaporate off the chlorine.
- Fresh cow manure
- Fresh cow urine if available (collected with a bucket held behind a cow!)
- Dried lentils or beans (*no soybeans*, but use lentils, mung, chickpeas, black beans, adzuki, kidney, pinto, etc.)
- Sugar (unbleached and less processed is better) or high sugar fresh fruits (sapotes, bananas, watermelon, dates, plums, or coconut water/sugar cane juice).
- Handful of fresh compost or forest humus

The process is as follows:

1. In a grinder or food processor (Vitamix blender works great) blend 4.4 lbs. (2 kg) of lentils into a very smooth flour. Do it in two or three batches so as not to overheat the blender.
2. Collect 22 lbs. (10 kg) of fresh cow manure. This should be from grass fed cows *only*. A packed 5 gallon (19 liter) bucket full is about 22 lbs.
3. Fill a 55-gallon drum or equivalent *about 75% full* with water. You must leave room for the other ingredients, so do not overfill.
4. To the water add 1.5–2.5 gallons (5–9 liters) of fresh cow urine.
5. Now add the cow dung. Crumble it with your hands into small pieces and stir thoroughly. A thick branch or long stick of bamboo works well.
6. Next add 4.75 lbs. (2.2 kg) of sugar or crushed fruit/juices.
7. Now add 4.4 lbs. (2 kg) of lentil/bean flour. Stir thoroughly and mash up any clumps that form.
8. Next add the finished compost. A few handfuls will work. Stir thoroughly.

9. Then, fill up the barrel with water to about 90% capacity. Leave 5–6 inches at the top *unfilled* because the mix will ferment and expand slightly and you don't want it overflowing. Stir.
10. Finally, cover with a permeable barrier such as burlap, cloth, or window screen. This allows gases like methane to escape, and helps keep animals and insects out. *Stir the mixture* 3× *per day*. In warm temps (70–90°F or 21–32°C) it's ready in 3–4 days. In cooler temps allow 4–5 days to ferment. Make sure to vigorously stir 3× every day.
11. The *Jivamritam* is now ready. You now have enough innoculant for 1 acre. It can be carefully filtered through micron-sized cloth filters and then used in fertigation systems, or applied using buckets (although that's very labor intensive). It likely would make an effective foliar supplement as well. Make sure it is very well strained!

The originators of this process claim it *only works properly if the crops are very thickly mulched*. They also claim no other fertilizer is needed beyond *Jivamritam* if the crop is mulched. We've seen good success using it, but were not only using *Jivamritam*. Your success may vary, but it is worth knowing how to make it and possibly integrating it into your fertilizing program. Perhaps you may find that with mulch, it's all you need!

Notes

Chapter 1: Why Small Scale, Small Fruits?

1. According to Vedic sources, *Satya Yuga* is a period in the eternal rotating cycle of time wherein a renewed Earth is once again in harmony, balance and great abundance, as well as the human psyche and condition. *Satya Yuga* lasts 1,728,000 years, and we are about 420,000 years away from the Satya Yuga Cycle beginning again.
2. As the current *Kali Yuga* progresses, quality, yield, nutrition, and availability of food continually decreases.
3. Ambrosia Beetle Pests of Nursery and Landscape Trees, NC State Extension Publications, content.ces.ncsu.edu/granulate-asian-ambrosia-beetle-1

Chapter 2: 21st Century Planning

1. An incredible resource is the USDA soil survey map program. It can tell you the soil type of every location in the USA! It can be found here: websoilsurvey.sc.egov.usda.gov/App/WebSoilSurvey.aspx
2. Cheryl Kaiser, Matt Earnst, "High Tunnel Tomatoes," University of Kentucky, College of Agriculture, Food and Environment. www.uky.edu/ccd/sites/www.uky.edu.ccd/files/hightunneltomatoes.pdf
3. "GM Crops List," 2021, International Service for the Acquisition of Agri-biotech Applications (ISAAA), isaaa.org/gmapprovaldatabase/cropslist/default.asp
4. Kathy K. Garvey, "First Native American Honey Bee," July 2009, UC Davis Dept. of Entomology and Nematology, ucanr.edu/blogs/blogcore/postdetail.cfm?postnum=1544

Chapter 3: Sourcing Plants and Navigating the 21st Century Nursery Scene

1. gardenwatchdog.com is a great place to start the search for nursery reviews. Google also has reviews for most nurseries.
2. In my experience, bare-root plants dug in late fall can be held in cold storage until mid-April before the quality rapidly declines.

Chapter 4: Creating Beds

1. "Cereal Rye," Sustainable Agriculture Research and Education, 2007. sare.org/publications/managing-cover-crops-profitably/nonlegume-cover-crops/cereal-rye

Chapter 5: Managing Soil Fertility Organically

1. Carol Burns, Kenneth Bodnar, et al., "Cancer Incidence of 2,4-D Production Workers," *Int J Environ Res Public Health*, 2011 Sept. ncbi.nlm.nih.gov/pmc/articles/PMC3194105/
2. See: *One Straw Revolution and The Natural Way of Farming: The Principles and Practices of Green Philosophy* by Masanobu Fukuoka.
3. This process is credited to Subhash Palekar, a Hindu man in India who pioneered the process and teaches it to others, and has several books available on his 'Spiritual Farming' techniques. We created a video on how to make Jivamritam, linked here: youtube.com/watch?v=ejj8qn9PGLM

Chapter 6: Planting Successfully

1. Johnny's Selected seeds sells one here: johnnyseeds.com/tools-supplies/bed-preparation-tools/seedbed-rollers/johnnys-seedbed-roller-30%22-7219.html
2. There is some debate over whether topsoil should be returned first or the subsoil put in first. It likely does not make a big difference either way.

Chapter 7: Organic Weed Management

1. The following section on mulching is adapted from *Pawpaw: The Complete Growing and Marketing Guide*, New Society Publishers, 2021.
2. See BCS Plastic Mulch Layer. bcsamerica.com/product/plastic-mulch-layer

Chapter 9: Maintenance, Protection, and Organic Pest Control

1. Search Integrated Pest Management on Wikipedia. en.wikipedia.org/wiki/Integrated_pest_management
2. Eduardo Faundez, David Rider, "The brown marmorated stink bug…in Chile," Arquivos Entomolóxicos, April 20, 2017. researchgate.net/publication/316277383_The_brown_marmorated_stink_bug_Halyomorpha_halys_Stal_1855_Heteroptera_Pentatomidae_in_Chile
3. en.wikipedia.org/wiki/Spotted_lanternfly
4. "Spotted Lanternfly Alert," Pennsylvania Dept. of Agriculture. agriculture.pa.gov/Plants_Land_Water/PlantIndustry/Entomology/spotted_lanternfly/SpottedLanternflyAlert/Pages/default.aspx
5. "How to Build a…Lanternfly Circle Trap," Penn State Extension.

extension.psu.edu/how-to-build-a-new-style-spotted-lanternfly-circle-trap youtube.com/watch?v=yjOKIOOw1ZA

6. "Monitoring SWD with traps," Univ. of New Hampshire Extension. extension .unh.edu/resource/monitoring-spotted-wing-drosophila-swd-traps
7. "SWD and Backyard Small Fruit Production." Univ. of Kentucky College of Agriculture. entomology.ca.uky.edu/ef231
8. "Biologically-based Management of Arthropod Pests..." USDA Agricultural Research Service. ars.usda.gov/research/programs-projects/project/?accnNo =429790&fy=2019
9. "Fencing to Exclude Deer," North Carolina Wildlife Resources Commission. ncwildlife.org/Learning/Species/Mammals/Whitetail-Deer/Fencing-to-Exclude -Deer#42041180-permanent-fencing
10. "Adding a Third Dimension to wildlife barrier fence," Manitoba Cooperator. manitobacooperator.ca/livestock/adding-a-third-dimension-to-a-wildlife -barrier-fence/
11. Brian Miller, Georgia O'Malley, et al., "Electric Fences for Preventing Browse Damage by White-tailed Deer," Purdue Extension, FNR 136. extension.purdue .edu/extmedia/fnr/fnr-136.pdf
12. "Protecting vineyards from birds with drones," Fruit Growers News. fruitgrowersnews.com/news/protecting-vineyards-from-birds-with-drones -is-study-topic/
13. Ishara Rijal , "Use of Water Mist to Reduce the Risk of Water Damage," d.lib .msu.edu/etd/4634/datastream/OBJ/download/USE_OF_WATER_MIST_TO _REDUCE_THE_RISK_OF_FROST_DAMAGE_IN_TREE_FRUITS.pdf
14. Lucas McCartney, Mark Lefsrud, "Portable Frost-protection Misting System," American Society for Horticultural Science, June 2015. journals.ashs.org /horttech/view/journals/horttech/25/3/article-p313.xml
15. Joe Hannan, "Frost Protection for High Density Orchards," Iowa State University Extension. extension.iastate.edu/smallfarms/frost-protection-high -density-orchards

Chapter 10: Blackberries

1. R.T. Jones, J.G. Strang, "Growing Raspberries and Blackberries in Kentucky," U of K, College of Agriculture, Cooperative Extension Service. ca.uky.edu /agcomm/pubs/ho/ho15/ho15.pdf. Page 9
2. Ibid. Page 9
3. Ibid. Page 5
4. Ibid. Page 7
5. "Rednecked Cane Borer," Virginia Tech, Department of Entomology. virginiafruit.ento.vt.edu/redneck.html

Chapter 11: Blueberries

1. Ryan Pankau, "The history of blueberries..." Illinois Extension, Jan. 2019 extension.illinois.edu/blogs/garden-scoop/2019-01-19-history-blueberries-native-american-staple-domesticated-superfood
2. Leonard Perry, "Choosing Blueberries," Univ. of Vermont Extension pss.uvm.edu/ppp/articles/chooseblue.html
3. Mark Longstroth, Michigan State University Extension, canr.msu.edu/news/how_to_avoid_a_common_blueberry_planting_error
4. David Yarborough, "Home Garden Lowbush Blueberry Planting Guide," Univ. of Maine Extension, Feb. 2003. extension.umaine.edu/blueberries/factsheets/cultivated-lowbush-blueberries/home-garden-lowbush-blueberry-planting-guide/
5. Joe Masabni, "Blueberry Production," UKREC, Princeton, KY. uky.edu/Ag/Horticulture/masabni/PPT/blueberry.pdf
6. Eric Hanson, Mark Longstroth, "Using your sprinkler to protect blueberries from freezes," MSU Extension, April, 2007. canr.msu.edu/news/using_your_sprinkler_system_to_protect_blueberries_from_freezes
7. Some cultivar recommendations taken from *The Berry Growers Companion* by Barbra Bowling, 2000, Timber Press

Chapter 12: Raspberries

1. "Health and Healing Fact Sheets," Oregon State Extension Service. berryhealth.fst.oregonstate.edu/health_healing/fact_sheets/black_raspberry_facts.htm
2. Jean English, "Raspberries: Challenging But Profitable," Maine Organic Farmers and Gardeners, Spring, 2005. mofga.org/Publications/The-Maine-Organic-Farmer-Gardener/Spring-2005/Raspberries
3. "Stake Your Tomatoes," Penn State Extension Service, August 2016. extension.psu.edu/stake-your-tomatoes
4. "Trellis Systems," NC State Extension Publications. content.ces.ncsu.edu/southeast-regional-caneberry-production-guide/trellis-systems
5. Raspberry (*Rubus* spp)—Root Rot," Pacific Northwest Pest Management Handbooks. pnwhandbooks.org/plantdisease/host-disease/raspberry-rubus-spp-root-rot
   Michael Ellis, "Phytophthora Root Rot of Raspberry," Ohio State University Extension Ohioline. ohioline.osu.edu/factsheet/plpath-fru-14

Chapter 14: Juneberries

1. Jim Ochterski, "Juneberries—They Go Where Blueberries Can't," Cornell CALS, Oct. 2011. smallfarms.cornell.edu/2011/10/juneberries-they-go-where-blueberries-cant/

2. *Saskatoon Berry Production Manual*, Alberta, Agriculture and Forestry. agric.gov.ab.ca/app08/showpublications
3. Jim Ochterski, "Juneberries."
4. Kevin Laughlin, Robert Askew, Ronald Smith, "Juneberry," NDSU Extension Service. fruit.wisc.edu/wp-content/uploads/sites/36/2016/03/juneberry.pdf
5. Cheryl Kaiser, Matt Ernst, "Juneberries," U of K Extension Service, CCD-CP-11. uky.edu/ccd/sites/www.uky.edu.ccd/files/juneberry.pdf
6. "Saskatoons and Juneberries, What They Are..." https://pickyourown.org/saskatoons.php
7. Richard G. St-Pierre, "Growing Saskatoons, A Manual for Orchardists. pickyourown.org/saskatoons/Growing_Saskatoons_-_A_Manual_For_Orchardists.pdf
8. Scroll down at this Q & A site: wildflower.org/expert/show.php?id=10721
9. Sources: fruit.wisc.edu/wp-content/uploads/sites/36/2016/03/juneberry.pdf
   uky.edu/ccd/sites/www.uky.edu.ccd/files/juneberry.pdf
   honeyberryusa.com/honeyberry-plants-4.html

Chapter 15: Muscadine Grapes

1. Mark Hoffman, Patrick Conner, et al., *Muscadine Grape Production Guide for the Southeast*, NC State Extension. site.extension.uga.edu/viticulture/files/2020/05/Muscadine-Hort-Guide.pdf
   Commercial Muscadine and Bunch Grape Production Guide, Alabama Coop Extension. ssl.acesag.auburn.edu/pubs/docs/A/ANR-0774/ANR-0774-archive.pdf
2. famu.edu/viticulture/FAMU_Viticulture_Production%20Guide%20for%20Muscadines(3).pdf
3. Mary Ann Lila, "Why Wild Plants Can Protect You From Cancer." doctoroz.com/article/why-wild-plants-can-protect-you-cancer

Chapter 16: Mulberries

1. "Red Mulberry," Massachusetts Natural Heritage & Endangered Species Program. mass.gov/files/documents/2016/08/qd/morus-rubra.pdf
2. "Popcorn Disease on Mulberry," Texas Agrilife Extension Service. aggie-horticulture.tamu.edu/galveston/Gardening_Handbook/PDF-files/GH-067--popcorn-disease-on-mulberry.pdf
3. Tanglefoot® is an extremely sticky petroleum-derived glue-like substance that almost any insect sticks to and cannot pass over. It is OMRI listed.
4. In Asian countries they have already accomplished this, although the colder climate of the USA may require different genetics. bangkokpost.com/life/social-and-lifestyle/404722/a-luscious-looking-remedy
5. In 2021 EKU (Eastern Kentucky University) started a mulberry trial based on our recommendation, using 'Illinois Everbearing' as the cultivar, and trialing various training systems.

Chapter 17: Gooseberries

1. Kirk Pomper, "Gooseberries and Currants," Horticulture, Kentucky State University. kysu.edu/wp-content/uploads/2017/08/pompergoosetalk07.pdf
2. "Insect Pests in Currants and Gooseberries," Penn State Extension. extension.psu.edu/insect-pests-in-currants-and-gooseberries-in-home-fruit-plantings

Chapter 18: Currants

1. canr.msu.edu/chestnuts/resources/midwest-nut-producers-council/index
2. "Disease Descriptions for Gooseberries and Currants," Penn State Extension. extension.psu.edu/disease-descriptions-for-gooseberries-and-currants

Chapter 19: Figs

1. In the Bhagavad Gita (10:26) Krishna says, "Of all trees I am the holy fig tree (Asvattha) and amongst sages and demigods I am Narada. Of the singers of the gods [Gandharvas] I am Citraratha, and among perfected beings I am the sage Kapila." Asvattha (Ficus benghalensis) is also known as the "boddhi" tree and is the tree Guatama received enlightenment under, becoming the "Buddha."
2. High tunnels are simply wide, long greenhouses made of arching half-circle steel hoops covered in very thick, durable transparent plastic. Their internal climate is drier and hotter than outside.
3. "Climate Battery Greenhouse, Version 1," Threefold Farm. threefoldfarm.org/climate-battery-greenhouse
4. "...Building a Walipini Greenhouse," Mother Earth News. motherearthnews.com/organic-gardening/essential-tips-for-building-a-durable-walipini-greenhouse-zbcz1706
5. "Integrated Pest Management (IPM) for Gophers, Birds and Deer," UC Master Gardeners of Monterey and Santa Cruz, March, 2020. ucanr.edu/sites/MontereyBayMasterGardeners/files/321464.pdf
6. USDA 2014
7. Ernst, University of Kentucky, 2018

Chapter 20: Tomatoes

1. Wild Boar Farms. wildboarfarms.com/
2. "Stake Your Tomatoes," PennState Extension, August, 2016. extension.psu.edu/stake-your-tomatoes
3. "Blossom end rot of tomato tip sheet," Michigan State Extension, Sept. 2016. canr.msu.edu/resources/blossom_end_rot_tip_sheet

4. Michael Grieneilson, Brenna Aegerter, et al., "Growers' Guide to Grafted Tomatoes," Dept. of Land, Air, and Water Resources, UC, Davis. ucanr.edu/sites/veg_crop_sjc/files/264559.pdf

Chapter 21: Other Small Fruits Worthy of Consideration...Or Not

1. American Aronia Berry. midwestaronia.org/
2. Leslie Mertz, "Tarts with a Touch of Romance," Good Fruit Grower, Univ. of Saskatchewan. goodfruit.com/tarts-with-a-touch-of-romance/
3. David Brinn, "Israel's Elderberry Remedy..." Israel 21C, October 2004 israel21c.org/israels-elderberry-remedy-sambucol-provides-solution-to-u-s-flu-vaccine-shortage/
4. "The Grape Man of Texas," Grayson College, grayson.edu/munson/index.html

Chapter 22: Harvesting and Post-Harvesting

1. Prana is a Sanskrit concept meaning essential life force energy.
2. Steep Hill Equipment Solutions. steephillequipment.com/equipment-category/berry-equipment/

Chapter 23: Market Planning and Strategies

1. Berry Hill Drip Irrigation Specialists, under Packaging. berryhilldrip.com

Chapter 24: The Future of Small Fruit Growing

1. agron-www.agron.iastate.edu/courses/Agron541/classes/541/lesson04a/4a.7.html
2. Karli Petrovic, "How Peach Growers Can Overcome Low Chill Hours," Growing Produce, Feb. 2018. growingproduce.com/fruits/stone-fruit/peach-growers-can-overcome-low-chill-hours/
3. npgsweb.ars-grin.gov/gringlobal/Search

Resources

A.M Leonard: Quality hand tools, spades, hori hori, hoes, Gripple™ trellis components, etc. www.amleo.com

Earth Tools: KY-based mail-order company dealing in the highest quality European and Japanese hand tools and small farming equipment, including BCS and Grillo tractors. www.earthtools.net

Peaceful Heritage Nursery: Certified Organic resilient fruiting plants, fruit trees, berry plants, figs, fruit cuttings, scion wood, and seeds. Specializing in small fruits, pawpaws, mulberries, figs, and exotics. www.peacefulheritage.com

Farmtek: Mail order source for landscape fabric, frost blankets, and thousands of other farm supplies. www.farmtek.com

Farm Plastic Supply: Silage tarps and other products. www.farmplasticsupply.com/

Berry Hill Irrigation: Mail order irrigation supplies, and a source for clamshell berry containers and packaging. www.berryhilldrip.com

Nourse Farms: Wholesale, small sized, high-quality berry starts. www.noursefarms.com

www.growingfruit.org: Online forum with tons of information and where you can find local/regional growers to connect with.

eBay.com: Good source of organic pesticide products, fertilizers, and random supplies.

Seven Springs Farm: Mail order organic fertilizers, pesticides, and other products. Good pricing and selection. www.springsfarm.com

Orchard Valley Supply: High quality trellis supplies, and other farm supplies. www.orchardvalleysupply.com

Steep Hill Equipment: Professional berry cleaning equipment. www.steephillequipment.com/equipment-category/berry-equipment/

NAFEX: North American Fruit Explorers is a fun group that tries to locate, propagate and save hardy and unusual fruit cultivars and species. Members can share scion wood, cuttings, and information. www.nafex.org

USDA Germplasm Repository: This USDA branch will ship researchers and professional growers rare and valuable fruit plant cuttings for free through the GRIN system. www.ars.usda.gov/pacific-west-area/corvallis-or/national-clonal-germplasm-repository/docs/npgs-repositories/

Johnny's Selected Seeds: Mail-order source of small farming tools, frost blankets, bender for making metal hoops out of EMT tubes, tapeners, cover crop seeds, berry rake, hori hori, kama, and the best wearable harvest tubs. www.johnnyseeds.com
Welter Seed: Mail-order source of cover crop seed at good prices. www.welterseed.com
Coolbot: This device can be purchased direct at www.storeitcold.com
Plantra: Bird protection and other planting supplies. www.plantra.com
Deerfencing.com: Quality deer fencing supplies shipped to your door. www.deerfencing.com
Boot Strap Farmer: Mail orders source for hoop benders, shade cloth, cell trays, and lots of useful supplies. www.bootstrapfarmer.com
Beneficial Insects by Mail Order: www.insectary.com and www.rinconvitova.com
Drip Depot: Fertigation pumps and irrigation supplies by mail order. Large selection. www.dripdepot.com

Recommended Reading

The Living Soil Handbook by Jesse Frost. Chelsea Green Publishing.
The Berry Growers Companion by Barbra L. Bowling. Timber Press.
The Holistic Orchard: Tree Fruits and Berries the Biological Way by Michael Phillips. Chelsea Green Publishing.
Growing Organic Orchard Fruits by Danny L. Barney. Storey Publishers.
Rodale's Successful Organic Gardening: Fruits and Berries by Lee Reich. Rodale Press.
Uncommon Fruits for Every Garden by Lee Reich. Timber Press.
Growing Figs in Cold Climates by Lee Reich. New Society Publishers.
Managing Cover Crops Profitably. SARE Handbook 9.
How to Grow More Vegetables by John Jeavons. Rodale Press.

Index

About the Author

Blake Cothron is an organic farmer, nurseryman, writer, musician, and speaker, with over 20 years experience in organic agriculture, botany, horticulture, and growing food. He is the co-owner and co-operator of Peaceful Heritage Nursery, a 4-acre USDA Certified Organic research farm, orchard, and edible plant nursery. Author of *Pawpaws: The Complete Growing and Marketing Guide*, he has also written for *Permaculture Design Magazine* and various online publications. Blake has an educational blog and YouTube channel devoted to fruit growing, and he is an educator with the Organic Association of Kentucky. He divides his time between farming, research, writing, beekeeping, gardening, travel, yoga, meditation, and being a husband and father in beautiful Kentucky.

ABOUT NEW SOCIETY PUBLISHERS

New Society Publishers is an activist, solutions-oriented publisher focused on publishing books to build a more just and sustainable future. Our books offer tips, tools, and insights from leading experts in a wide range of areas.

We're proud to hold to the highest environmental and social standards of any publisher in North America. When you buy New Society books, you are part of the solution!

At New Society Publishers, we care deeply about *what* we publish—but also about *how* we do business.

- All our books are printed on 100% **post-consumer recycled paper**, processed chlorine-free, with low-VOC vegetable-based inks (since 2002). We print all our books in North America (never overseas)
- Our corporate structure is an innovative employee shareholder agreement, so we're one-third employee-owned (since 2015)
- We've created a Statement of Ethics (2021). The intent of this Statement is to act as a framework to guide our actions and facilitate feedback for continuous improvement of our work
- We're carbon-neutral (since 2006)
- We're certified as a B Corporation (since 2016)
- We're Signatories to the UN's Sustainable Development Goals (SDG) Publishers Compact (2020–2030, the Decade of Action)

To download our full catalog, sign up for our quarterly newsletter, and to learn more about New Society Publishers, please visit newsociety.com

ENVIRONMENTAL BENEFITS STATEMENT

New Society Publishers saved the following resources by printing the pages of this book on chlorine free paper made with 100% post-consumer waste.

TREES	WATER	ENERGY	SOLID WASTE	GREENHOUSE GASES
104 FULLY GROWN	8,300 GALLONS	43 MILLION BTUs	360 POUNDS	44,900 POUNDS

Environmental impact estimates were made using the Environmental Paper Network Paper Calculator 4.0. For more information visit www.papercalculator.org